CAP 413

# RADIOTELEPHONY
# PROCEDURES AND PHRASEOLOGY

*Civil Aviation Authority London September 1978*

©Civil Aviation Authority 1978

First published September 1978
Reprinted January 1979

ISBN 0 86039 068 3

Printed in England by Ebenezer Baylis and Son Ltd.,
and distributed by Civil Aviation Authority, Greville House,
37 Gratton Road, Cheltenham

## FOREWORD

This book, which replaces CAP 46 and Aeronautical Information Circular 101/72, is based on the International Standards and Recommended Practices for Aeronautical Communications contained in ICAO Annex 10 Volume 2 (Communications Procedures) to the Convention on International Civil Aviation, the UK Air Pilot (COM/MET/RAC/SAR sections) and the CAA Manual of Air Traffic Services.

It is the standard reference for the examination in the procedures section of the Flight Radiotelephony Operators (General and Restricted) Licence, details of which are contained in CAP 90.

Chapters 1, 2, 3 and 4 relate to radio operating procedures while chapters 5, 6, 7, 8 and 9 contain standard pilot/air traffic service radiotelephony phraseology for private and commercial flying. Separate type faces are used to indicate pilot or ground originated messages.

In order to minimise amendments, the callsigns and frequencies used will only be changed when a new edition is published. Current operational details are to be found in the United Kingdom Air Pilot.

All enquiries about the text are to be addressed to:

The Secretary
RTF Phraseology Working Group
Civil Aviation Authority
Room T 1224 CAA House
45-59 Kingsway
London WC2B 6TE

CONTENTS

# GLOSSARY

## DEFINITIONS

### AERONAUTICAL BROADCASTING SERVICE.

A broadcasting service intended for the transmission of information relating to air navigation.

### AERONAUTICAL MOBILE SERVICE.

A radiocommunication service between aircraft stations and aeronautical stations, or between aircraft stations.

### AERONAUTICAL RADIONAVIGATION SERVICE.

A radiodetermination service for the benefit of aircraft intended for the determination of position or direction, or for obstruction warning in navigation.

### AERONAUTICAL STATION.

A land station in the aeronautical mobile service. In certain circumstances an aeronautical station may be placed on board a ship or an earth satellite.

### AERONAUTICAL TELECOMMUNICATION SERVICE.

A telecommunication service provided for any aeronautical purpose.

### AIR-GROUND COMMUNICATION.

Two way communication between aircraft and stations or locations on the surface of the earth.

### AIR-TO-GROUND COMMUNICATION.

One way communication from aircraft to stations or locations on the surface of the earth.

### AIRCRAFT STATION.

A mobile station in the aeronautical mobile service on board an aircraft.

AIR TRAFFIC SERVICE.

A generic term meaning variously, flight information service, alerting service, air traffic advisory service, air traffic control service.

AIR TRAFFIC ADVISORY SERVICE.

A service provided within advisory airspace to ensure separation, in so far as possible, between aircraft which are operating on IFR flight plans.

AIR TRAFFIC CONTROL SERVICE.

A service provided for the purpose of;

(1) Preventing collisions:
   (a) between aircraft and
   (b) on the manoeuvring area between aircraft and obstructions and,
(2) Expediting and maintaining an orderly flow of air traffic.

ALERTING SERVICE.

A service provided to notify appropriate organisations regarding aircraft in need of search and rescue aid and to assist such organisations as required.

FLIGHT INFORMATION SERVICE.

A service provided for the purpose of giving information useful for the safe and efficient conduct of flights.

BLIND TRANSMISSION.

A transmission from one station to another station in circumstances where two-way communication cannot be established, but where it is believed that the called station is able to receive the transmission.

BROADCAST.

A transmission of information relating to air navigation that is not addressed to a specific station or stations.

DISTRESS.

A condition of being threatened by serious and/or imminent danger and of requiring immediate assistance.

GROUND-TO-AIR COMMUNICATION.

One way communication from stations or locations on the surface of the earth to aircraft.

RADIO BEARING.

The angle between the apparent direction of a definite source of emission of electro magnetic waves and a reference direction, as determined at a radio direction finding station. A TRUE radio bearing is one for which the reference direction is that of true North. A MAGNETIC radio bearing is one for which the reference direction is that of magnetic North.

RADIO DIRECTION FINDING STATION.

A radio station intended to determine only the direction of other stations by means of transmissions from the latter.

SECONDARY SURVEILLANCE RADAR.

A system of radar using ground interrogators and airborne transponders to determine the position of aircraft in range and azimuth and, when the agreed modes and codes are used, height and identity as well.

TELECOMMUNICATIONS.

Any transmission, emission or reception of signs, signals, writing, images and sounds or intelligence of any nature by wire, radio, visual or other electro-magnetic systems.

URGENCY.

A condition concerning the safety of an aircraft or other vehicles or of some person on board or within sight, but which does not require immediate assistance.

## ABBREVIATIONS

| *Abbreviation* | *Meaning* |
|---|---|
| ACFT | Aircraft. |
| AFIS | Aerodrome Flight Information Service. |
| AIP | Aeronautical Information Publication. |
| ANO | Air Navigation Order. |
| APP | Approach Control. |
| ATIS | Automatic Terminal Information Service. |
| ATSU | Air Traffic Service Unit. |
| AWY | Airway. |
| C/S | Callsign. |
| CAA | Civil Aviation Authority. |
| CTR | Control Zone. |
| *DME | Distance Measuring Equipment. |
| *ETA | Estimated Time of Arrival. |
| FIS | Flight Information Service. |
| GMC | Ground Movement Control. |
| *GMT | Greenwich Mean Time. |
| *IFR | Instrument Flight Rules. |
| *ILS | Instrument Landing System. |
| *IMC | Instrument Meteorological Conditions. |
| *NDB | Non-directional radio beacon. |
| NM | Nautical Miles. |
| *QFE | Atmospheric pressure at aerodrome elevation (or at RWY threshold). |
| *QNH | Altimeter sub-scale setting to obtain elevation when on ground. |
| RTF | Radiotelephone. |
| RWY | Runway. |

| *Abbreviation* | *Meaning* |
|---|---|
| SSR | Secondary Surveillance Radar. |
| TWR | Aerodrome Control. |
| VDF | Very high frequency direction finding station. |
| *VFR | Visual Flight Rules. |
| *VMC | Visual Meteorological Conditions. |
| *VOR | VHF Omni-directional radio range. |

* Indicates that the abbreviation is normally spelt out without using the phonetic alphabet of Chapter 2 para 2.

CHAPTER 1

# GENERAL INSTRUCTIONS

## 1 LANGUAGE TO BE USED

The English language is used internationally for air-ground radiotelephony communication.

## 2 TIME SYSTEM

Greenwich Mean Time (GMT) is to be used by all stations.

Midnight is designated as 2400 hours for the end of the day and 0000 hours for the beginning of the day.

Radio time signals are transmitted at frequent intervals by the British Broadcasting Corporation and details are published in the AIP (COM section). The correct time may be obtained by RTF from any ATSU.

## 3 HOURS OF SERVICE AND COMMUNICATIONS WATCH

The hours of service of the radio facilities available in the United Kingdom are published in the AIP (COM 2 section), which also details those periods set aside for maintenance.

Aircraft in flight are to maintain watch as required by the Competent Authority (CAA) and are not to cease watch, except for reasons of safety, without informing the aeronautical station concerned and stating the time at which it is expected watch will be resumed. If it is necessary to suspend watch beyond this time a revised time should, if possible, be notified before the original time of expiry.

## 4 RECORD OF COMMUNICATIONS

An aircraft telecommunication log-book is not required to be kept in respect of communication by radiotelephony with a radio station on land or on a ship which provides a radio service for aircraft. Most UK aeronautical stations have automatic recording equipment for air-ground communications but in those that do not a record is kept of all communications with aircraft.

5 **CATEGORIES OF MESSAGE**

The categories of messages handled by the aeronautical mobile service are in the following order of priority:

| | |
|---|---|
| (a) Distress messages and distress traffic.<br>(b) Urgency messages. | See Chap. 3 |
| (c) Communications relating to direction finding. | See Chap. 4. |
| (d) Flight safety messages.<br>(e) Meteorological messages.<br>(f) Flight Regularity messages. | See Chap. 10 and App A for detailed contents. |

CHAPTER 2

# BASIC OPERATING PROCEDURES

## 1 TRANSMITTING TECHNIQUES

### 1.1 Introduction

The articulation of users of radiotelephony communication can affect the efficient reception of speech. Poor delivery of RTF can cause confusion or misunderstanding and create potentially hazardous situations.

### 1.2 Before Transmitting

Listen out on the frequency to be used to ensure that you will not interfere with any transmission from another station. Do not interrupt if you hear a station making a transmission that obviously requires a reply from another addressee. However, any station having a distress or urgency message to transmit is entitled to interrupt any transmission of lower priority (see Chapter 1 para. 5).

Determine what you wish to say before you transmit and avoid the inadvertent use of 'hesitation sounds' like 'er', 'ah', or 'um' etc.

### 1.3 Microphones

Microphones of different types have a wide range of operating characteristics. Ensure that you position the microphone in accordance with its operating instructions.

Do not turn your head away from the microphone while talking or vary the distance between it and your mouth. Severe distortion of speech may arise from:

(a) Talking too close to the microphone.
(b) Touching the microphone with the lips.
(c) Holding the microphone or boom (of a combined headset/ microphone system).

Make certain the transmit switch is actuated before you start talking and not released until you have finished. This will prevent 'clipping' of transmissions.

The Air Navigation Order (Operation of radio in aircraft) restricts the use of hand held microphones.

## 1.4 Pronunciation

Pronounce each word clearly and end it correctly. Do not run words together or mispronounce/omit consonants. Regional variations of vocabulary, syntax and accent may make mutual understanding difficult or impossible over a radiotelephone.

## 1.5 Voice Control

Maintain the speaking volume at a constant level—do not whisper but, more important, do not shout.

Avoid lowering your voice at the end of each transmission. Maintain an even rate of speech (not greater than 100 words a minute).

Do not talk rapidly.

Talk slightly slower if any details of a message have to be written down by the recipient.

Numbers are more easily understood if a slight pause is made before and after the number.

## 1.6 Phraseology

Use standard phraseology where possible.
Always be concise and unambiguous on RTF.
Avoid asking questions by using an inflection of the voice.
Always use the correct interrogative word or standard phrase.
Do not use expressions of politeness to excess.

# 2 PRONUNCIATION OF LETTERS AND THE MORSE CODE

2.1 The following words are to be used when it is required to transmit individual letters, e.g. callsigns.

| Letter | Word | Pronunciation in English | Morse Code |
| --- | --- | --- | --- |
| A | Alfa | <u>AL</u> FAH | · – |
| B | Bravo | <u>BRAH</u> <u>VOH</u> | – · · · |
| C | Charlie | <u>CHAR</u> LEE | – · – · |
| D | Delta | <u>DELL</u> TAH | – · · |
| E | Echo | <u>ECK</u> OH | · |
| F | Foxtrot | <u>FOKS</u> TROT | · · – · |
| G | Golf | GOLF | – – · |
| H | Hotel | HOH <u>TELL</u> | · · · · |
| I | India | <u>IN</u> DEE AH | · · |
| J | Juliett | <u>JEW</u> LEE <u>ETT</u> | · – – – |
| K | Kilo | <u>KEY</u> LOH | – · – |
| L | Lima | <u>LEE</u> MAH | · – · · |
| M | Mike | MIKE | – – |
| N | November | NO <u>VEM</u> BER | – · |
| O | Oscar | <u>OSS</u> CAH | – – – |
| P | Papa | PAH <u>PAH</u> | · – – · |
| Q | Quebec | KEY <u>BECK</u> | – – · – |
| R | Romeo | <u>ROW</u> ME OH | · – · |
| S | Sierra | SEE <u>AIR</u> RAH | · · · |
| T | Tango | <u>TANG</u> GO | – |
| U | Uniform | <u>YOU</u> NEE FORM | · · – |
| V | Victor | <u>VIK</u> TAH | · · · – |
| W | Whiskey | <u>WISS</u> KEY | · – – |
| X | Xray | <u>ECKS</u> <u>RAY</u> | – · · – |
| Y | Yankee | <u>YANG</u> KEY | – · – – |
| Z | Zulu | <u>ZOO</u> LOO | – – · · |

The underlined syllables are emphasised.

2.2 Some abbreviations have become unmistakeable through common usage and may be transmitted without using the phonetic word for each letter e.g. ILS, QNH, QFE, QDM.

## 3 PRONUNCIATION OF DIGITS

The following words are used when transmitting single digits.

| | |
|---|---|
| 0—ZERO | 5—FIFE |
| 1—WUN | 6—SIX |
| 2—TOO | 7—SEVEN |
| 3—TREE | 8—AIT |
| 4—FOWER | 9—NINER |

The underlined syllables are emphasised.

*5 and 9 are spoken as 'FIFE' and 'NINER' respectively as they are easily confused when spoken normally.*

## 4 TRANSMISSION AND VERIFICATION OF NUMBERS

### 4.1 Transmission of Numbers

Every digit of a number is normally spoken separately. However, if the number is a whole thousand, each digit of the number of thousands is spoken separately, followed by the word 'thousand'. (Pronounced 'TOUSAND'.)

A decimal point within a number is to be indicated by the word 'decimal' (pronounced DAY-SEE-MAL).

| NUMBER | TRANSMITTED AS: |
|---|---|
| 10 | WUN ZERO |
| 75 | SEVEN FIFE |
| 100 | WUN ZERO ZERO |
| 295 | TOO NINER FIFE |
| 1000 | WUN TOUSAND |
| 1012 | WUN ZERO WUN TOO |
| 25000 | TOO FIFE TOUSAND |
| 118.3 | WUN WUN AIT DAY-SEE-MAL TREE |

4.2 **Verification of Numbers**

When it is desired to verify the accurate reception of numbers the person transmitting shall request the receiving operator to repeat all the numbers.

## 5 TRANSMISSION OF TIME

When transmitting time by radiotelephony, only the minutes of the hour are normally required. However, if there is any possibility of misunderstanding, the hour is to be included.

Each digit is spoken separately with the pronunciation shown in 2.3.

| TIME | TRANSMITTED AS: |
|---|---|
| 0920 | TOO ZERO<br>*or*<br>ZERO NINER TOO ZERO |
| 1650 | FIFE ZERO<br>*or*<br>WUN SIX FIFE ZERO |

## 6 COMMUNICATIONS STANDARD WORDS AND PHRASES

The words and phrases shown in the table below are to be used whenever applicable. Where the phrase is self explanatory no meaning is indicated.

| Word/Phrase | Meaning |
|---|---|
| 'Acknowledge' | Let me know that you have received and understood this message. |
| 'Affirmative' | Yes *or* permission granted. |

| Word/Phrase | Meaning |
|---|---|
| 'Break' | I hereby indicate the separation between portions of the message. |
| 'Correction' | An error has been made in this transmission (or message indicated). The correct version is . . . |
| 'How do you read?' | What is the readability of my transmission. |
| 'I say again' | *Self explanatory* |
| 'Negative' | No, *or* permission not granted, *or* that is not correct. |
| 'Over' | My transmission is ended and I expect a response from you. |
| 'Out' | This conversation is ended and no response is expected. |
| 'Pass your message' | *Self explanatory* |
| 'Read back' | Repeat all, or the specified part, of this message back to me exactly as received. |
| 'Roger' | I have received all of your last transmission. |
| 'Say again' | Repeat all, or the following part, of your last transmission. |
| 'Speak slower' | *Self explanatory* |
| 'Standby' | *Self explanatory* |
| 'That is correct' | *Self explanatory* |

| Word/Phrase | Meaning |
|---|---|
| 'Verify' | Check coding, check text with the originator and send correct version. |
| 'Wilco' | Your last message (or message indicated) received, understood and will be complied with. |
| 'Words Twice' | As a request: Communication is difficult. Please send every word twice.<br><br>As information: Since communication is difficult every word in this message will be sent twice. |

## 7 RADIOTELEPHONY CALLSIGNS

### 7.1 Air Traffic Service Units

The location of the unit is used followed by a suffix to indicate the service provided.

| Callsign Suffix | Service |
|---|---|
| Tower | Aerodrome Control |
| Approach | Approach Control |
| Zone | Zone Control |
| Control | Area Control |
| Ground | Ground Movement Control |
| Radar/Director | Approach Radar |
| Information | Flight Information Service |
| Radio | Air-ground service only |

Full details of all ATSU callsigns are in the AIP (COM 2).

## 7.2 Aircraft

(a) *Full Callsigns*
Aircraft are identified on RTF by one of the following types of callsign:

| Type | Example |
|---|---|
| 1. The registration of the aircraft. | GBFBO: N58746: 7TVEL |
| 2. The registration of the aircraft preceded by the telephony designator of the aircraft operating agency. | MINAIR GBFBO |
| 3. The flight identification or trip number. | SPEEDBIRD 628 |

The full callsign is always to be used when establishing communication (see para 9).

An aircraft is not to change the type of callsign during flight.

The suffix 'HEAVY' after the callsign indicates the wake vortex category of the aircraft.

(b) *Abbreviation of callsigns*

After communication has been established, and the aeronautical station is satisfied that no confusion will arise, it may abbreviate the aircraft callsign as shown in the following table.

A pilot may only abbreviate the callsign of his aircraft if it has first been abbreviated by the aeronautical station.

| Type | Full Callsign | Abbreviated Callsign |
|---|---|---|
| 1 | GBFBO<br>N58746<br>7 TVEL | G BO<br>N 746<br>7 EL |
| 2 | MINAIR GBFBO | MINAIR BO |
| 3 | SPEEDBIRD 628 | No abbreviated form |

(c) *Special callsigns*

Special callsigns may be used by certain aircraft for specific State purposes e.g.

TESTER 22
RESCUE 41 (Search and Rescue)

(d) *Callsign confusion*

The similarity of callsigns can cause confusion and misunderstanding. The ATSU will normally warn the pilot when this is likely and suggest a remedy for the problem.

i. Completely different callsigns may be identical when abbreviated.
    GATSB = G SB
    GBRSB = G SB
  Both pilots will be instructed to use their full callsigns.
ii. Radiotelephony designators may be similar or trip numbers identical.
    DANAIR
    FINAIR
    SWISSAIR
  Pilots will be warned by the ATSU where trip numbers are identical and the company designator should be emphasised in all messages.
iii. Trip numbers may contain digits in common.
    Caledonian 982
    Caledonian 902
iv. If trip numbers are 360 or less, confusion can be caused by the similarity of these numbers with flight levels or radar headings.

## 8 COMPOSITION OF MESSAGES

Messages normally consist of a Call and a Text:

| | |
|---|---|
| *Call* | **'Stansted Tower this is GBFBO** |
| *Text* | **Taxi Instructions Over'** |

However, VHF radiotelephony communication lends itself in certain circumstances to abbreviation and omission of some items (see paras 7.2.(b) and 10 and Chapter 5).

## 9 ESTABLISHMENT OF COMMUNICATION

Aircraft shall, if possible, communicate directly with the aeronautical station appropriate to the area in which they are flying.

### 9.1 Initial Call

Callsign of station being called.
The words 'this is'.
Callsign of calling station.
Invitation to reply.

| *ACFT* | **'Stansted Tower this is GBFBO over'** |
|---|---|

After a call has been made to an aeronautical station a pilot should wait at least 10 seconds before making a second call. This will help reduce unnecessary transmissions while the aeronautical station is preparing to reply to the initial call.

### 9.2 Reply to an Initial Call

Callsign of station calling.
Callsign of station answering.
Invitation to proceed with the transmission.

| *TWR* | 'GBFBO Stansted Tower—Pass your Message' |
|---|---|

### 9.3 Additional Requirements and Information

Where an ATSU has more than one frequency allocated in the AIP (COM section) for a particular Air Traffic Service, the initial call should be followed by an indication of the frequency used.

When using HF equipment and provided there is no possibility of confusion, identification of the transmitting frequency may be achieved by stating only the first two digits: i.e. Speedbird 325 calling Shanwick Radio on 8910 KHz.

| *ACFT* | **'Shanwick Radio Speedbird 325 on 89 Over'** |
|---|---|

If a station hears a call without being certain that it is the station being called it should not reply until the call has been repeated and is understood.

When a station is called but is uncertain of the identification of the calling station it shall reply by transmitting:

| | |
|---|---|
| *TWR* | 'Station Calling Stansted Tower—Say again your Callsign' |

If an aeronautical station is called simultaneously by more than one aircraft it shall decide the order in which the aircraft shall communicate.

## 10 SUBSEQUENT RADIOTELEPHONY COMMUNICATION

### 10.1 Shortening of Procedure

(a) If it is certain that the station called will receive the call then the calling station may transmit the message without repeating the Initial Call and Reply procedure.

| | |
|---|---|
| *ACFT* | **'Stansted Tower this is GBFBO Downwind Over'** |

(b) After contact has been established certain phrases may be omitted provided no confusion is likely to arise: 'THIS IS', 'OVER', 'ROGER', etc.

| | |
|---|---|
| *ACFT* | **'Stansted Tower GBFBO Downwind'** |

(c) Provided no mistake in identity is likely to occur continuous two-way communication is permitted without further identification or call until termination of the contact.

| | |
|---|---|
| *ACFT* | **'G-BO Downwind'** |
| *TWR* | 'Clear to Final—Report Base Leg' |
| *ACFT* | **'Wilco'** |

10.2 **Acknowledgement of Receipt**

Acknowledgement of receipt of a message transmitted by either (i) an aircraft or (ii) an aeronautical station is the callsign of the aircraft:

(i)

| | |
|---|---|
| *ACFT* | **'G-BO Clear of the Runway'** |
| *TWR* | 'G-BO' |

(ii)

| | |
|---|---|
| *TWR* | 'G-BO Flock of Birds at 2 Miles Final' |
| *ACFT* | **'G-BO'** |

Certain ATC instructions require a full or partial readback as well as the acknowledgement of receipt.

10.3 **Readback**

Readback by pilots of any ATC messages listed in (a) or (b) should be in the same order or sequence as transmitted.

(a) The following ATC instructions are to be read back in full by the pilot.

The readback is to be concluded with the aircraft callsign.

| |
|---|
| LEVEL INSTRUCTIONS |
| HEADING INSTRUCTIONS |
| AIRWAYS OR ROUTE CLEARANCES |
| CLEARANCE TO ENTER OR CROSS AN ACTIVE RUNWAY |
| SSR OPERATING INSTRUCTIONS |
| ALTIMETER SETTINGS |
| VDF INFORMATION |
| FREQUENCY CHANGES |

| | |
|---|---|
| *TWR* | 'G-BO QFE 1014' |
| *ACFT* | **'QFE 1014 G-BO'** |

Pilot acknowledgement of any of these items without a comprehensive readback should not be accepted by a controller. If the controller does not receive a readback he should ask the pilot to read back the instruction as issued.

Similarly, the pilot is expected to request that the above items are repeated or clarified if any are not fully understood.

(b) Other ATC executive instructions should be read back in an abbreviated form and concluded with the aircraft callsign.

| | |
|---|---|
| *TWR* | 'G-BO Hold Position. Aircraft on Final for Runway 23' |
| *ACFT* | **'Holding G-BO'** |

(c) Acknowledgement of ATC information is by the aircraft callsign.

(d) If the readback of instructions is incorrect the station shall transmit the word 'NEGATIVE' followed by the correct version:

Example: Incorrect read back of altimeter setting.

| | |
|---|---|
| *ACFT* | **'QNH 1014 G-BO'** |
| *TWR* | 'Negative, QNH 1024' |

(e) It is essential that careful attention is paid both to instructions and their readback since it is very easy to hear only what is expected to be heard.

### 10.4 Corrections

When an error has been made during transmission the word CORRECTION shall be spoken, the last correct phrase repeated, followed by the correct version.

| | |
|---|---|
| *TWR* | 'G-BO—Caution—Marked Trench Right Side Correction Marked Trench Left Side of Entrance to Apron'. |

10.5 **Repetitions**

(a) If reception conditions are considered to be difficult transmit the important elements of the message twice.

(b) If the receiving operator is in doubt as to the correctness of the message received he shall request repetition either in full or in part.

i. Repetition of an entire message:

'SAY AGAIN'

ii. Repetition of a specific item:

'SAY AGAIN QNH'

iii. Repetition of part of a message:

'SAY AGAIN ALL BEFORE . . .
(the first word satisfactorily received)

'SAY AGAIN ALL AFTER . . .'
(the last word satisfactorily received)

## 11 FAILURE TO ESTABLISH OR MAINTAIN COMMUNICATION

11.1 **Air to Ground**

(a) Check the following points:

i. The correct frequency has been selected for the route being flown.
ii. The Aeronautical Station being called is open for watch.
iii. The aircraft is not out of radio range.

(b) If the previous points are in order it may be that the aircraft equipment is not functioning correctly—complete the checks of headset and radio installation appropriate to the aircraft.

(c) If the pilot is still unable to establish communication on any designated aeronautical station frequency, or with any other aircraft, the pilot is to transmit his message twice on the designated frequency preceded by the phrase 'TRANSMITTING BLIND' in case the transmitter is still functioning.

(d) Where a transmitter failure is suspected, check or change the microphone. Listen out on the designated frequency for instructions. It should be possible to answer questions by use of the carrier wave if the microphone is not functioning.

(e) In the case of a receiver failure transmit reports at the scheduled times or positions on the designated frequency preceded by the phrase 'TRANSMITTING BLIND DUE TO RECEIVER FAILURE'.

Transmit the message twice and include the time of next intended transmission and any change of frequency.

(f) An aircraft which is being provided with air traffic control or advisory service is to transmit information regarding the intention of the pilot in command with respect to the continuation of the flight.

Specific procedures for the action to be taken by pilots of IFR and Special VFR flights are contained in the appropriate AIP (RAC) sections.

## 11.2 Ground to Air

After completing checks of ground equipment (most airports have standby and emergency communications equipment) the ground station will request other aeronautical stations and aircraft to attempt to communicate with the aircraft concerned.

If still unable to establish communication the aeronautical station will transmit messages addressed to the aircraft by blind transmission on the frequency on which the aircraft is believed to be listening.

These will consist of:

(a) The level, route and EAT (or ETA) to which it is assumed the aircraft is adhering and:

(b) The weather conditions at the destination aerodrome and suitable alternate and, if practicable, the weather conditions in an area or areas suitable for descent through cloud procedure to be effected. (See AIP RAC Section.)

## 11.3 Relaying of Messages

(a) If an aircraft is unable to communicate directly with the appropriate air-ground station it shall use any relay means available to transmit messages to that station.

(b) When an aircraft originates a message that requires retransmission over the Aeronautical Fixed Service the message shall comprise of the following parts in the order shown:
Call
Address (preceded by the word 'FOR')
Text

| | |
|---|---|
| *ACFT* | **'Stansted Tower this is GBFBO for Connaught Operations Newcastle. Returning 2300 Request Extension of Hours'** |

## 12 FREQUENCIES TO BE USED

### 12.1 Aircraft Operation

Aircraft shall operate on the appropriate radio frequencies designated by the aeronautical station or as promulgated in the AIP (COM section).

### 12.2 Transfer of Communication

An aircraft should be advised by the appropriate aeronautical station to transfer from one radio frequency to another. In the absence of such advice the aircraft should notify the aeronautical station before such transfer takes place. An aircraft which has transferred communication watch from one radio frequency to another shall establish communication as shown in 2.9.

The instruction to change frequency will be passed in the following form:

(a) the identity of the unit to be contacted.
(b) the frequency to be used for contact.

| | |
|---|---|
| *TWR* | 'G-BO at time 07 Contact London Control 129.6 Over'<br>*or*<br>'G-BO Contact London Control 129.6 Over' |
| *ACFT* | **'London Control 129.6 G-BO'** |

If no further communication is received from the pilot after correct readback of the frequency, transfer of communication will be assumed.

If a pilot fails to establish communication after transfer he should revert to the original frequency and ask for instructions.

## 13 TEST PROCEDURES

### 13.1 Aircraft Equipment

The consent of the aeronautical station shall be obtained before an aircraft, for test or adjustment purposes, makes any transmission which may interfere with the aeronautical station. If a station in the aeronautical mobile service makes test signals, either for the adjustment of a transmitter before making a call, or the adjustment of a receiver, such signals shall not continue for more than 10 seconds and shall be composed of spoken digits (ONE, TWO, THREE etc) followed by the callsign of the station transmitting the test signals.

### 13.2 Aircraft Radio Checks with Air Traffic Services

Aircraft may make three types of general radio check with an ATSU—each one indicates the nature of the check being effected:

**'SIGNAL CHECK'**—indicates the aircraft is airborne

**'MAINTENANCE CHECK'**—indicates a routine ground test

**'PREFLIGHT CHECK'**—indicates aircraft is about to depart.

The standard phrase 'HOW DO YOU READ' may also be used by pilots or controllers.

In all instances the response should indicate the readability of the signal by selection of the appropriate figure from the following scale:

| Scale | Meaning |
|---|---|
| 1 | Unreadable |
| 2 | Readable now and then |
| 3 | Readable but with difficulty |
| 4 | Readable |
| 5 | Perfectly readable |

## 13.3 Phraseology

(a) *Aircraft request for test*

(i) The identification of the station being called.
(ii) The words 'THIS IS'.
(iii) The aircraft identification.
(iv) The words 'SIGNAL/PREFLIGHT/MAINTENANCE CHECK as appropriate.
(v) The frequency being used.
(vi) The word 'OVER'

| *ACFT* | **'Stansted Tower this is GBFBO Preflight Check 118.15 Over'** |
|---|---|

(b) *ATC response*

(i) The identification of the aircraft.
(ii) The identification of the ground station replying.
(iii) Readability of the aircraft transmission.
(iv) The word 'OUT'.

| *TWR* | 'GBFBO Stansted Tower Readability Three Out' |
|---|---|

## 13.4 Additional Tests

(a) *Local*

An ATSU may, on behalf of the local telecommunications officer, ask pilots to make tests of a radio or radar aid.

(b) *Aircraft-Approval and Licensing Procedure*
Details of the procedures for 'Radio Test Flights' concerned with the Approval and Licensing of aircraft radio stations are to be found in the AIP (COM section).

(c) *Ground equipment transmissions*
Any transmission using the identification 'TST' (-/· · ·/- in morse) is radiating for test purposes only and must not be used for operational purposes.

# CHAPTER 3

# DISTRESS AND URGENCY—RADIO TELEPHONY COMMUNICATION PROCEDURES

## 1 INTRODUCTION

The general procedures detailed in this chapter should be used in the Aeronautical Mobile Service. Further information is available in the AIP (SAR).

Pilots should use their radios to request assistance whenever they are in doubt as to the safe conduct of a flight. RTF transmissions should be made slowly and distinctly, each word being clearly pronounced to assist transcription.

## 2 DISTRESS RADIOTELEPHONY COMMUNICATION

### 2.1 General

An aircraft in distress may use any available means of attracting attention, making known its position and obtaining assistance. Distress traffic has absolute priority over other transmissions. The first transmission of the distress message shall be on the frequency in use at the time and continued on that frequency until it is considered that better assistance can be provided by transferring. It is important that prior to any change of frequency the aircraft shall transmit on the frequency on which the distress message was sent an appropriate message indicating the frequency to which it intends to change.

If the aircraft was not previously in communication with an ATSU, or is unable to re-establish communication, then the distress message should be transmitted on the Emergency Frequency 121.5 MHz (or any other available frequency). If necessary and time permits, an aircraft equipped to communicate on the maritime distress frequency of 2182 KHz should transmit the distress message on that frequency in an effort to alert ships and coast stations which may be within range. Certain North Atlantic Ocean Station vessels keep a listening watch only on 121.5 MHz and all aircraft crossing the North Atlantic are required to listen out on 121.5 MHz.

All stations hearing distress traffic must immediately cease any transmission likely to interfere with the distress traffic and listen on the frequency used for the distress traffic.

A distress message should normally be addressed to the station with which the aircraft was last in communication or in whose area of responsibility the aircraft is operating, but it may be broadcast if time and circumstances make this preferable. The station addressed will assume control of the aircraft in distress. If the distress message is not acknowledged by the station addressed, any station hearing the distress message shall immediately acknowledge it and assume control of the communications or transfer the responsibility, advising the aircraft if a transfer is to be made. (See para 2.5.) Any aircraft or aeronautical station which has knowledge of distress traffic but which cannot itself assist the station in distress should monitor such traffic until it is evident that assistance is being provided.

## 2.2 The Distress Signal

The radiotelephony distress signal consists of the word 'MAYDAY'. Morse code 'S O S' (. . . - - - . . .). Any call or message relating to the immediate assistance required by an aircraft or vessel in distress should be prefixed 'MAYDAY'.

## 2.3 Distress Message Transmitted by an aircraft in distress

A distress message when transmitted by an aircraft in distress should consist of the distress signal 'MAYDAY', preferably spoken 3 times, followed by as many as possible of the following elements, spoken distinctly, in the order shown:

(a) Name of station addressed (when applicable)
(b) Callsign and type of aircraft
(c) Nature of Emergency
(d) Intention of the person in command
(e) Present position, flight level/altitude, heading.

| | |
|---|---|
| *ACFT* | **"MAYDAY MAYDAY MAYDAY<br>Stansted Tower this is GBFBO Cessna 310 starboard engine on fire require immediate landing at Stansted 5 miles south of Stansted flight level 65 heading 300"** |

Note: The procedure described above should be followed if time and circumstances permit. However, under conditions which do not allow the procedure to be followed in full, an aircraft may use any means at its disposal to attract attention and make known its condition, including the activation of the appropriate SSR mode and code in those areas in which SSR is available (Code 7700).

## 2.4 Distress Message Transmitted by an aircraft not itself in distress

Any aeronautical station or aircraft having knowledge of a distress incident may transmit a distress message whenever it is considered that such action is necessary to obtain assistance for the aircraft or vessel in distress. In such circumstances it should be made clear that the station transmitting is not itself in distress.

| | |
|---|---|
| *ACFT* | **'MAYDAY MAYDAY MAYDAY**<br>**Stansted Tower this is GBFBO. Have intercepted MAYDAY from GBFVD I say again from GBFVD Cessna 172 engine failed ditching 10 miles east of Clacton 1000 feet descending heading 260 Over'** |

## 2.5 Acknowledgement of Distress

Acknowledgement of a distress message shall take the following form:

| | |
|---|---|
| *ATSU* | 'GBFBO Stansted Tower Roger MAYDAY Out' |

## 2.6 Imposition of Silence

The aircraft in distress or the station in control of distress traffic may impose silence either on all stations in the area or on any particular station that interferes with the distress traffic.

It shall address these instructions to 'ALL STATIONS' or to one station only according to circumstances.

In either case the words to be used are:

'STOP TRANSMITTING MAYDAY'

| | |
|---|---|
| *ATSU* | 'All Stations (or ACFT callsign) Stop transmitting MAYDAY Stansted Tower Out' |

2.7 **Transfer of Other Aircraft to another Frequency**

The aeronautical station acknowledging the distress message may consider it prudent to transfer other aircraft to another frequency in order to give the distress traffic a discrete frequency.

| | |
|---|---|
| *TWR* | 'MAYDAY GBFBO all other aircraft contact Stansted Tower on 123.8' |

2.8 **Emergency Descent**
It may be necessary for an aircraft to carry out an emergency descent to a lower level or for landing.

| | |
|---|---|
| *ATSU* | 'Emergency to all Concerned<br><br>Emergency Descent at . . . (aerodrome/ holding facility/location)<br><br>All Aircraft Below . . . (flight level/feet)<br><br>Within . . . Miles of . . . (aerodrome/holding facility/location)<br><br>Leave . . . (location/locality) Immediately |

Where standard routes for leaving the area are not published, routeing instructions will be given according to the circumstances.

2.9 **Cancellation of Distress Communications and of Silence Condition**

When an aircraft is no longer in distress it shall transmit a message cancelling the distress condition.

| | |
|---|---|
| *ACFT* | **'MAYDAY—Stansted Tower this is GBF VD Cancel Distress. Engine started. Proceeding to Southend'** |

When distress traffic has ceased or when silence is no longer necessary on the frequency used for the distress traffic, the station which has been controlling this traffic shall transmit on that frequency a message addressed to 'all stations' indicating that normal working may be resumed. The message will include the words:

'Distress Traffic Ended'

| *ATSU* | 'MAYDAY all Stations this is Stansted Tower. 1030 Hours GBFVD. Distress Traffic Ended Out' |
|---|---|

## 3 URGENCY RADIOTELEPHONY COMMUNICATION

### 3.1 General

URGENCY communications have priority over all other communications except distress and all stations shall take care not to interfere with the transmission of urgency traffic. The urgency message to be sent by an aircraft reporting an urgency condition should normally be addressed to the station with which the aircraft was last in communication or in whose area of responsibility the aircraft is operating.

### 3.2 Urgency Signal

The radiotelephony urgency signal consists of the word 'PAN' which should also be used (once) when transmitting messages—other than the first urgency message—relating to the urgency condition. (The morse code for PAN is — · · — / — · — / — · —)

### 3.3 Urgency Message

The urgency message transmitted by the aircraft should consist of the urgency signal 'PAN' spoken 3 times followed by as many as possible of the details below and in the order shown:

(a) The name of the station addressed.
(b) The callsign and type of aircraft.
(c) Nature of urgency.

(d) Intention of the person in command.
(e) Present position, flight level/altitude and heading.
(f) Qualification of the pilot in one of these forms:
  (i) Student pilot.
  (ii) No instrument qualifications.
  (iii) IMC Rating.
  (iv) Full Instrument Rating.

| | |
|---|---|
| *ACFT* | **'PAN PAN PAN**<br>**Stansted Tower this is GBFBO Cessna 310 My Passenger is Ill. Diverting to Southend. 10 Miles East of Clacton at 2500 feet heading 200. Full Instrument Rating'** |

## 4 THE UK EMERGENCY SERVICE FOR PILOTS IN DISTRESS, URGENCY OR DIFFICULTY.

### 4.1 Introduction

The Emergency Service operates on the Emergency Frequency 121.5 MHz and provides a continuously available communications and aid service from a number of civil and military aerodromes and other ATSUs, details of which are shown in the AIP (COM section).

The Distress and Diversion elements at London and Scottish Air Traffic Control Centres can also use certain VDF and radar equipment to provide an emergency aid and fixer service. However, the obtaining of a VDF position fix takes time because it entails manual plotting of bearings received by telephone from those VDF stations selected by the Emergency Controller. The height of the aircraft and its position relative to the VDF station(s) determines the accuracy of VDF bearings and the position fix. Below 3000 feet amsl the service is extremely limited.

Once an aircraft position has been fixed it may then be possible to use radar to identify and assist it.

The selection of SSR Mode A Code 7700 (where possible) greatly assists identification of the aircraft.

In military aviation 'TYRO' is used before the callsign to indicate that the pilot is inexperienced. Civil pilots, if in communication with a military unit, or the Emergency Controller at an ATCC, are invited to use 'TYRO' should they feel uncertain of their ability to cope with a situation.

## 4.2 Use of the Emergency Service

(a) *Distress and Urgency*

When the Emergency Frequency 121.5 MHz is to be used by a pilot it is not necessary to address the call to a specific centre or ATSU. The Emergency Controller at the London or Scottish Centre will normally answer the call.

Once communication has been established on 121.5 MHz, pilots should not leave that frequency without the agreement of the Emergency Controller.

(b) *Difficulty*

The Emergency Service is also available to pilots in difficulty or in need of assistance (eg when uncertain of position) as distinct from 'DISTRESS' or 'URGENCY'.

Pilots in difficulty should use the Emergency Service at the appropriate ATCC (RTF callsigns are 'DRAYTON CENTRE' or 'SCOTTISH CENTRE') on 121.5 MHz.

The content of the message should be as shown in para 3.3 excluding the URGENCY SIGNAL.

# CHAPTER 4

# RADIOTELEPHONY DIRECTION FINDING

## 1 INTRODUCTION

The aeronautical stations that offer a VDF service are listed in the AIP (COM section).

Some VDF stations stipulate that the service is not available for en-route navigation purposes (except in emergency).

VDF bearing information will only be given when conditions are satisfactory and radio bearings fall within the calibrated limits of the station. If the provision of a radio bearing is not possible the pilot will be told the reason.

A VDF station will give the following as requested:

(a) Magnetic bearing of the aircraft from the station (using the signal 'QDR' or appropriate phrase).

(b) Magnetic heading to be steered by the aircraft (assuming no wind) to reach the VDF station (using the signal 'QDM' or appropriate phrase).

(c) True bearing of the aircraft from the station (using the signal 'QTE' or appropriate phrase).

(d) True track to be steered by the aircraft (assuming no wind) to reach the VDF station (using the signal 'QUJ' or appropriate phrase. Rarely used.).

The majority of VDF stations in the UK can only provide QDRs or QDMs (as in a. and b.) so a request for a QTE may, at times, elicit a QDR instead.

## 2 REQUESTS FOR VDF SERVICE

To request a bearing or heading to steer, the aircraft calls the aeronautical station on the designated or listening frequency and then specifies required service by means of the appropriate phrase or 'Q' signal. The transmission is to be ended with the aircraft callsign.

VDF stations will normally provide the appropriate bearing immediately without a further transmission, however the aircraft may have to wait until adjustments and checks are made to the equipment. If a further transmission is required the VDF station will request the pilot to transmit for DF. VDF equipment requires a minimum transmission time of 3 seconds to provide a bearing indication.

If a VDF station is not satisfied with its observation it will request the aircraft to repeat its transmission.

3 **REPORTING VDF INFORMATION**

If the aircraft has requested a bearing or heading to steer the VDF station will advise the aircraft of the results in the following form:

(a) The appropriate phrase or 'Q' signal.
(b) The bearing or heading to steer sent as three digits.
(c) Class of bearing.
(d) Time of observation (if necessary).

The 'Class of bearing' referred to in (c) indicates the estimated accuracy of the bearing taken by the VDF station and the table below shows the limits of any fluctuations observed.

| |
|---|
| CLASS A—accurate to within ± 2 degrees |
| CLASS B—accurate to within ± 5 degrees |
| CLASS C—accurate to within ± 10 degrees |
| CLASS D—accuracy worse than Class C |

4 **VDF RTF PHRASEOLOGY**

| | |
|---|---|
| *ACFT* | **"Stansted Tower this is GBFBO. Request QDM. GBFBO Over."** |
| *ATSU* | "GBFBO this is Stansted Tower. QDM 040 degrees Class Bravo." |
| *ACFT* | **"QDM 040 degrees Class Bravo GBFBO"** |

*or*

| | |
|---|---|
| *ACFT* | **"Stansted Tower this is GBFBO. Request QDR. GBFBO Over"** |
| *ATSU* | "GBFBO this is Stansted Tower Standby" |

*then*

| | |
|---|---|
| *ATSU* | "G-BO Stansted—Transmit for QDR" |
| | **"Stansted G-BO. Request QDR G-BO"** |
| *ATSU* | "G-BO Stansted QDR 320 degrees Class Charlie" |
| *ACFT* | **"QDR 320 degrees Class Charlie G-BO"** |

Pilots are required to read back VDF Information.

CHAPTER 5

# GENERAL PHRASEOLOGY

## 1 INTRODUCTION

Phraseology is as easy to learn correctly as incorrectly. The use of non-standard phraseology or poor operating techniques inevitably results in ambiguity, misunderstanding, confusion or danger.

In the following chapters the callsigns used will be of two aircraft, GBFBO and GBFVD and two ATSUs, Stansted and London, so that examples given will be realistic.

The following words may be omitted from transmissions if no confusion or ambiguity will result:

'Surface wind' and 'Knots' when giving the surface wind direction and speed.

'Degrees' for surface wind direction and radar headings.

'Visibility', 'Cloud amount', 'Height' in meteorological reports.

'Over', 'Roger', 'This is', 'Out' etc.

Acknowledgement of receipt and readback of instructions (where not indicated) should be as detailed in chapter 2 para 10.3.

This chapter contains more detailed explanation of phraseology than later chapters in order to meet the needs of the trainee pilot or controller.

## 2 GROUND MOVEMENT AND TAKE OFF

### 2.1 General

The intention in this section is to progress through RTF transmissions in the order in which they are likely to be met.

The examples demonstrate exchanges between a pilot and a controller providing an aerodrome control service. Details of techniques and phraseology for use with the Aerodrome Flight Information Service are contained in para 5.6.

### 2.2 Preflight

(a) *Special VFR flights*

A flight plan is not normally required for SVFR but ATC must receive details of:

Aircraft callsign
Aircraft type
Pilot's intentions

These details may be passed by RTF or, at busy aerodromes, through the Flight Clearance office.

| | |
|---|---|
| *ACFT* | **'. . . Tower GBFBO Cessna 310 request special VFR to leave the zone to the north en route to Birmingham'** |

The TWR response may consist of taxi instructions as in para 2.3.

(b) *Signal checks*
Pilots will normally check their RTF equipment with Aerodrome Control using the procedures in chapter 2 para 13.

The Controller will pass, on request, meteorological details, runway in use, time checks, transition level etc.

(c) *Departure Approval (IFR flights)*
This procedure (normally for IFR flights on Airways) is detailed in Chapter 7.

(d) *Pushback*
At some aerodromes aircraft may be parked nose in to the terminal in order to save parking space, the aircraft have to be pushed back from the terminal.

| | |
|---|---|
| *ACFT* | **'. . . Tower GBFBO. Push back from Stand 26'** |
| *TWR* | 'G-BO cleared to push back.'<br>*or*<br>'G-BO—hold position.' |

(e) *Engine Start*
Requests to start engines are normally made by pilots of jet or propjet aircraft in order to avoid excessive fuel penalties caused by delays on the ground.

| | |
|---|---|
| *ACFT* | **'. . . Tower. G-BO request start'** |
| *TWR* | *'G-BO cleared to start temperature plus 8'*<br>*or*<br>'G-BO standby for start. Temperature plus 8' |
| *ACFT* | **'G-BO'** |

2.3 **Taxi Instructions**

(a) *Clearance Limit and Routeing*

The clearance limit issued by the controller is the point at which the aircraft must stop unless further permission to proceed is received. It will normally be the holding point of the runway in use but may be another position on the aerodrome if demanded by the traffic situation. The route to be followed will also be passed. The table shows a variety of TWR instructions:

| *ACFT* | **'. . . Tower—G-BO taxi instructions'** |
|---|---|
| *TWR* | 'G-BO cleared to the holding point runway 23 via taxiways 2 and 4. QNH 1012'.<br><br>'G-BO cleared via taxiway 2 to hold short of taxiway 5. QNH 1012'.<br><br>'G-BO due work in progress cleared via taxiway 5 to backtrack runway 05. QNH 1012'.<br><br>'G-BO hold position'.<br><br>'G-BO give way to the B707 entering the apron, cleared to the holding point runway 05 via taxiway 2. QNH 1012'.<br><br>'G-BO cleared to the holding point runway 05 via taxiway 2. Follow the B707 ahead. QNH 1012.'<br><br>'G-BO cleared to the holding point runway 05 via taxiway 2. QFE 1002' (QFE for circuit traffic)<br><br>'G-BO surface wind 320 degrees 26 knots. Runway 23 or 05 available.' |

See Chapter 2 para 10.3. for read back requirements.

(b) *Essential Aerodrome Information*

Information about the manoeuvring area or associated facilities that would affect the safe operation of aircraft.

| | |
|---|---|
| *TWR* | 'G-BO Caution—small marked trench right side of taxiway at holding point' |
| *ACFT* | **'G-BO'** |

(c) *Holding*

Since a collision may occur if an aircraft does not stop at a prescribed point, instructions to hold should be correctly passed and acknowledged.

| | |
|---|---|
| *TWR*<br><br>*ACFT* | *'G-BO hold position'*<br><br>**'Holding G-BO'** |
| *TWR*<br><br>*ACFT* | 'G-BO hold short of the apron'<br><br>**'Wilco G-BO'** |
| *TWR*<br><br>*ACFT* | 'G-BO hold south of runway 23'<br><br>**'Hold south of 23 G-BO'** |

(d) *Runway Crossing*

The pilot requests permission to cross a runway or the TWR controller anticipates the request and issues clearance before the aircraft reaches its clearance limit. The procedures in (c) *holding*, are applicable when clearance to cross cannot be given.

| | |
|---|---|
| *ACFT* | **'. . . Tower. G-BO may I cross runway 23'** |
| *TWR* | 'G-BO cross. Report clear'<br><br>'G-BO cross runway 23 at the intersection report clear'<br><br>'G-BO after the landing B707 cross runway 23. Report clear'<br><br>'G-BO after the landing B707 cross runway 23 at the intersection report clear'<br><br>'G-BO after the departing B707 cross runway 23. Report clear'<br><br>'G-BO after the departing B707 cross runway 23 at the ramp. Report clear'. |

2.4 **ATC Clearance**

The issue of an ATC route or VFR clearance is not permission to take-off and pilots must wait for the specific clearance to take-off (see para. 2.6) before commencing take-off run. The following are examples of ATC clearances—note readback requirements in Chapter 2 para 10.3.

(a) *VFR*

| | |
|---|---|
| *TWR* | 'G-BO after departure leave the zone to the north VFR' |
| *ACFT* | **'Leave the zone to the north. VFR G-BO'** |

(b) *SVFR (zone clearance)*

| | |
|---|---|
| *TWR* | 'G-BO after departure cleared to leave the zone special VFR to the north not above 2000 feet QNH 1014' |
| *ACFT* | **'Cleared to leave the zone special VFR to the north not above 2000 feet. QNH 1014. G-BO'** |

(c) *IFR (airways clearance)*

| | |
|---|---|
| *TWR* | 'G-BO is cleared to Manchester via Matching, Brookmans Park, Amber 2, Amber 1 east to maintain flight level 60' |
| *ACFT* | **'Cleared to Manchester via Matching, Brookmans Park, Amber 2, Amber 1 east maintain flight level 60. G-BO'** |

(d) *IFR (zone clearance)*

| | |
|---|---|
| *TWR* | 'G-BO after departure cleared to leave the zone to the north at 3000 feet' |
| *ACFT* | **'Cleared to leave the zone to the north at 3000 feet. G-BO'** |

At certain aerodromes the IFR Zone clearance may consist of a STANDARD INSTRUMENT DEPARTURE (SID), details of which are to be found in the AIP (RAC section).

In (c) and (d) radar directions and SSR instructions may be passed.

The phrase 'UNABLE TO CLEAR AT . . . (LEVEL/ROUTE)' may be used.

### 2.5 Holding Point Instructions

(a) *Transfer to Tower Control*

Busy units with GMC and Tower Control will transfer the aircraft to Tower Control at, or before, the holding point for take off instructions.

| | |
|---|---|
| *GMC* | 'G-BO contact tower 118.4' |
| *ACFT* | **'Tower 118.4 G-BO'** |

(b) *Controller able to issue take off clearance*

If the departure clearance is short and uncomplicated it may be included with the take off clearance.

| | |
|---|---|
| *ACFT* | **'. . . Tower G-BO ready for take off'** |
| *TWR* | 'G-BO left hand circuits. 270 degrees 10 knots. Cleared take off'<br><br>'G-BO leave the zone to the south VFR. 270 degrees 10 knots. Cleared take off'<br><br>'G-BO right turn on track Matching 080 degrees 15 knots. Cleared take off'<br><br>'G-BO 270 degrees 10 knots. Cleared take off report airborne'<br><br>'G-BO Viscount on 5 mile final cleared immediate take off. 270 degrees 10 knots' (immediate take off will not be given to wide body aircraft) |
| *ACFT* | **'Take off G-BO'** |

(c) *Controller not able to issue take off clearance*

| | |
|---|---|
| *ACFT* | **'G-BO ready for take off'** |
| *TWR* | 'G-BO hold position'<br><br>'G-BO line up and hold'<br><br>'G-BO after the landing Viscount (has passed you) line up and hold'<br><br>'G-BO after the departing Viscount (has passed you) line up and hold' |

The expression 'LINE UP AND HOLD' is used in the UK to instruct the pilot to enter the runway and hold at the take off position.

(d) *Aircraft lined up and waiting*
Take off clearance will take the form already shown in (b). The following instruction will only be given if the controller requires the runway to be vacated quickly.

| | |
|---|---|
| *TWR* | 'G-BO take off immediately or clear the runway—270 degrees 10 knots' |
| *ACFT* | **'Taking off G-BO'**<br>*or*<br>**'Leaving the runway G-BO'** |

2.6 **After Departure**

| | |
|---|---|
| *TWR*<br><br>*ACFT* | 'G-BO contact Stansted approach 126.95'<br><br>**'126.95 G-BO'** |
| *TWR*<br><br>*ACFT* | 'G-BO report passing the zone boundary'<br><br>**'Wilco'** |
| *TWR*<br><br>*ACFT* | 'G-BO report at 3000 feet'<br><br>**'Wilco'** |
| *TWR*<br><br>*ACFT* | 'G-BO report VMC on top'<br><br>**'Wilco'** |

3 **THE AERODROME TRAFFIC CIRCUIT**

3.1 **Reporting in the Aerodrome Traffic Circuit**

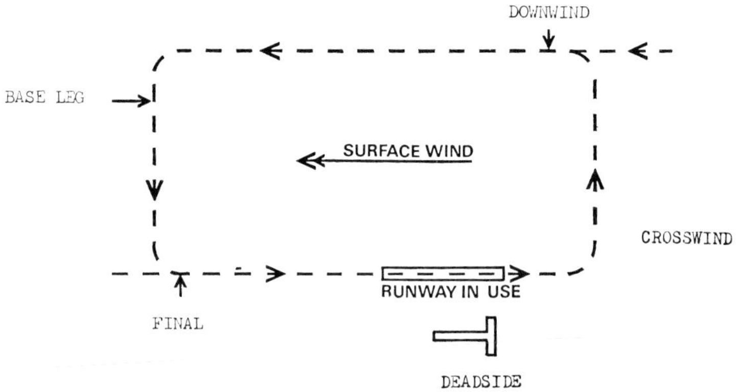


(a) *Crosswind*

A pilot is not normally required to report 'CROSSWIND'. However, a controller may pass instructions or information that contain references to 'CROSSWIND'.

| | |
|---|---|
| *TWR* | 'G-BO turn crosswind as soon as possible' |
| *ACFT* | **'Wilco G-BO'** |

(b) *Downwind*

This report should be made opposite the upwind end of the runway in use. It is helpful if, on the last circuit of a circuit training detail the pilot indicates his intention to make a full stop landing and not a touch and go.

(i)

| | |
|---|---|
| *ACFT* | **'G-BO downwind'**<br><br>**'G-BO downwind for landing'** |
| *TWR* | 'G-BO cleared to final' |

NB. The phrase 'CLEARED TO FINAL' indicates that G-BO is number one to approach and that no other aircraft is, or will be, between G-BO and the runway threshold.

(ii)

| *ACFT* | **'G-BO downwind'** |
|---|---|
| *TWR* | 'G-BO cleared to final number 3. Number 2 is a Jetstream on left base ahead' |

G-BO is number 3 to land. The controller should be advised if the traffic cannot be seen so that an alternative instruction may be given.

(iii)

| *ACFT* | **'G-BO downwind'** |
|---|---|
| *TWR* | 'G-BO cleared to final. Report before turning base' |

(iv)

| *ACFT* | **'G-BO downwind'** |
|---|---|
| *TWR* | 'G-BO cleared to final number 2 after the Tristar on base. You require 8 miles spacing—caution—Vortex Wake' |

(c) *Base Leg* (if requested)

| *ACFT* | **'G-BO base leg'** |
|---|---|
| *TWR* | 'G-BO cleared to final'<br><br>'G-BO cleared to final number 2 to the islander on final'<br><br>'G-BO cleared to final. One Islander to depart' |

| *ACFT* | **'G-BO ready to turn base'** |
|---|---|
| *TWR* | 'G-BO cleared to final'<br><br>'G-BO orbit right and report again on base'<br><br>'G-BO cleared to final number 2 to the Islander on final'<br><br>'G-BO continue downwind. I will tell you when to turn base' |

(d) *Long Final*

This report is made when aircraft turn on to final approach at a distance greater than 4nm from touchdown or when an aircraft on a straight in approach is 8nm from touchdown. In both cases the pilot is to report 'FINAL' at 4nm from touchdown.

| *ACFT* | **'. . . Tower G-BO long final'** |
|---|---|
| *TWR* | 'G-BO cleared to final' |
| | 'G-BO cleared to land runway 26. 280 degrees 10 knots' |

(e) *Final*

This report is normally made on completion of the turn on to final approach or, if the aircraft is making a straight in approach, approximately 4nm from the threshold. The controller should always state the runway with (i), (ii), (iii), or (iv).

| *ACFT* | | **'G-BO final'** |
|---|---|---|
| *TWR* | *(i)* | 'G-BO cleared to land runway 26. 280 degrees 10 knots' |
| | *(ii)* | 'G-BO cleared touch and go runway 26. 280 degrees 10 knots' |
| | *(iii)* | 'G-BO continue approach runway 26. HS748 departing' |
| | *(iv)* | 'G-BO land after the HS748 on runway 26. 280 degrees 10 knots' |
| | *(v)* | 'G-BO overshoot, I say again overshoot. Climb straight ahead for left hand circuit. Acknowledge' |

*Notes*

(iv) The 'Land after' procedure is described in the AIP (RAC section) 'Use of Runways' and the pilot is responsible for maintaining separation.

Certain very busy airports issue 'CLEAR TO LAND' when another aircraft is still on the runway. Strict criteria must be met by ATC—details are in the AIP (AGA 2 section).

(v) If an instruction to overshoot is given then it will also be amplified with the phrase 'NOT BELOW 400 FEET' if there is another aircraft on the runway.

If a pilot decides to overshoot then further instructions will be passed.

(f) *Taxi instructions after landing*

(i) On the runway.

| | |
|---|---|
| *TWR* | 'G-BO take the next convenient left then via taxiway two to the apron'<br><br>'G-BO continue to the end of the runway then right via taxiway 4 to the apron'<br><br>'G-BO cleared backtrack then right via taxiway 5 to stand 17' |

(ii) Clear of the runway

| | |
|---|---|
| *ACFT* | **'G-BO clear of runway 26'** |
| *TWR* | 'G-BO turn first left, then right to the apron' |

## 4 COMMON PHRASES

### 4.1 Level Instructions

(a) *To query the aircraft flight level, altitude or height.*

| | |
|---|---|
| *ATSU* | 'G-BO what is your level' |
| *ACFT* | **'G-BO maintaining flight level 40'**<br><br>**'G-BO maintaining 3000 feet QNH'**<br><br>**'G-BO maintaining 1500 feet QFE'** |

The word 'PASSING' may be substituted for 'MAINTAINING' when required.

(b) *Level information*

| | |
|---|---|
| *ATSU* | 'G-BO Report passing/leaving/reaching<br>Flight level 40'<br>3000 feet QNH'<br>1500 feet QFE' |
| *ACFT* | **'Wilco'** |

(c) *Level instructions given by an ATSU*

| |
|---|
| 'Maintain flight level 50'<br>'Maintain . . . feet QNH'<br>'Maintain . . . feet QFE'<br>'Maintain flight level 70 to Daventry' (reporting point)<br>'Maintain flight level 70 until past Daventry' (reporting point)<br>'Maintain flight level 70 while in controlled airspace'<br>'Maintain 1000 feet above/below GBFVD' (aircraft identity)<br>'Cross Daventry (reporting point) at flight level 70'<br>'Cross Daventry (reporting point) not above/not below flight level 80'<br>'Cross Daventry (reporting point) above/below flight level 80'<br>'Request level change en route to flight level 290' |

(d) *Level Change Instructions given by an ATSU*

> 'Climb/descend to flight level . . .'
>
> 'Climb/descend to 3000 feet QNH 1014' (mb)
>
> 'Descend to 1500 feet QFE 1010' (mb)

The pressure datum should always be stated—or requested—if the aircraft is operating below the transition level.

> 'After passing . . . (reporting point) climb/descend to flight level 40/3000 feet QNH 1014'
>
> 'Climb/descend at not more than/not less than . . . feet per minute'
>
> 'Climb/descend at . . . (time) to flight level 140'
>
> 'Climb/descend to reach flight level 120 at . . .' (time/reporting point)
>
> 'Climb/descend to flight level 50 maintaining VMC and own separation'
>
> 'Climb when instructed by radar to flight level 90'
>
> 'Climb/descend now to flight level 80'
>
> 'Expedite climb/descent to flight level 110'
>
> 'Climb/descend immediately to flight level 90'

The last phrase will only be used to resolve an urgent situation.

4.2 **Position Reporting**

(a) Unless otherwise instructed position reports shall be passed in the following manner:

Aircraft callsign
Position
Time over
Flight level or altitude
Next position and time over

| | |
|---|---|
| *ACFT* | **'London Control GBFBO Daventry at 47 flight level 80 estimating Lichfield at 0901'** |

(b) Omission of position reports.

| | |
|---|---|
| *ATSU* | 'Omit position reports while on this frequency'<br><br>'Resume position reporting'<br><br>'Next report at . . . ' (location) |

4.3 **Filing of flight plans when airborne**

Flight plans may be filed when airborne with any ATSU although it is normal to contact the FIR controller responsible for the area in which the aircraft is flying. If the airborne plan contains an intention to enter controlled airspace or certain Special Rules Areas/Zones then at least 10 minutes warning must be given. The message should start with the words 'I WISH TO FILE AN AIRBORNE FLIGHT PLAN' and should include the addresses to which it is to be sent. Details of the message will vary according to the type of flight (VFR/IFR) and pilots should follow the Flight Plan format.

4.4 **Passing information to an ATSU**

Certain services provided by an ATSU require information from the pilot before the service can be given: Flight Information Service, Airways joining, MATZ penetration, Aerodrome Flight Information Service, Lower Airspace Radar Service, Aerodrome Traffic Zone entry or transit.

Pilots should contact the service and state their intentions, in brief.

| *ACFT* | **'GBFBO To join at Daventry.'**<br>*or*<br>**To transit traffic zone.'** |
|---|---|

The ATSU will ask the pilot to pass his **details** and/or **route.** The following information should then be passed:

**DETAILS** Callsign
Aircraft Type
Position and Heading
Altitude or Flight Level
Flight Conditions

**ROUTE** Departure Aerodrome
Route including ETA for: Reporting Point
Joining Position
Holding or En-route facility
Zone Boundary
Destination aerodrome or point of first intended landing.
True Air Speed
Any other relevant information.

## 5 FLIGHT INFORMATION SERVICE

The service available is detailed in the AIP (RAC) section. Pilots wishing to use the FIS should establish communication and then pass both details and route.

| *ACFT* | **'London Information this is GBFBO'** |
|---|---|
| *ATSU* | 'GBFBO London Information pass your details and route' |
| *ACFT* | **'GBFBO Cessna 310**<br>**15 miles north of Stansted heading 320**<br>**Flight level 45 VMC**<br>**from Stansted estimating Daventry at 45 to Birmingham. 180 knots'** |

When a pilot is being given information which may affect the conduct of the flight the FIS will usually preface the message text with:

'You are informed that . . .'

| | |
|---|---|
| *FIS* | 'G-BO you are informed that severe icing has been reported at flight level 50 in the vicinity of Daventry' |

## 6 AERODROME FLIGHT INFORMATION SERVICE (AFIS)

### 6.1 Introduction

The AFIS provides information concerning traffic in the aerodrome traffic zone so that pilots can maintain their own separation from the other aircraft using the aerodrome.

### 6.2 Taxi Instructions

(a) *Clearance Limit*
The clearance limit is the point at which the aircraft must stop unless further permission to proceed is received. The taxi instructions will include the runway in use, circuit direction, the route to be followed and the clearance limit.

| *ACFT* | **'. . . Information—G-BO taxi instructions'** |
|---|---|
| *AFIS* | 'G-BO cleared to holding point runway 23, right hand circuit QNH 1012'<br><br>'G-BO cleared to holding point runway 23 via south taxiway. Left hand circuit. QNH 1012'<br><br>'G-BO hold position'<br><br>'G-BO cleared to holding point runway 23 follow the Cessna 172 ahead. Right hand Circuit QNH 1012' |

(b) *Essential Aerodrome Information*

Information about the manoeuvring area or associated facilities that would affect the safe operation of aircraft.

| | |
|---|---|
| *AFIS* | 'G-BO caution—small marked trench right side of taxiway at holding point' |
| *ACFT* | **'G-BO'** |

(c) *Holding*

Since a collision may occur if an aircraft does not stop at a prescribed point, instructions to hold should be correctly passed and acknowledged.

| | |
|---|---|
| *AFIS*<br><br>*ACFT* | 'G-BO hold position'<br><br>**'Holding G-BO'** |
| *AFIS*<br><br>*ACFT* | 'G-BO hold short of the apron'<br><br>**'Wilco G-BO'** |
| *AFIS*<br><br>*ACFT* | 'G-BO hold south of runway 23'<br><br>**'Hold south of 23 G-BO'** |

(d) *Runway Crossing*

The pilot requests permission to cross a runway or the AFIS Officer anticipates the request and issues clearance before the aircraft reaches its clearance limit. The procedures in (c) holding are applicable when clearance to cross can not be given.

| | |
|---|---|
| *ACFT* | **'G-BO may I cross runway 23'** |
| *AFIS* | 'G-BO cross. Report clear'<br><br>'G-BO cross runway 23 at the intersection. Report Clear'<br><br>'G-BO after the landing 150 cross runway 23. Report clear'<br><br>'G-BO after the departing 172 cross runway 23. Report Clear' |

(e) *When ready to take-off*

When an aircraft is ready for take-off, the AFIS Officer will only pass traffic information. The decision to take-off is left to the pilot.

| *ACFT* | **'G-BO ready for take-off'** |
|---|---|
| *AFIS* | 'G-BO take-off at your discretion one Cessna 172 turning base leg. 270 degrees 15 knots'<br><br>'G-BO line up and hold'<br><br>'G-BO hold position' |

6.3 **Aerodrome Traffic Circuit**

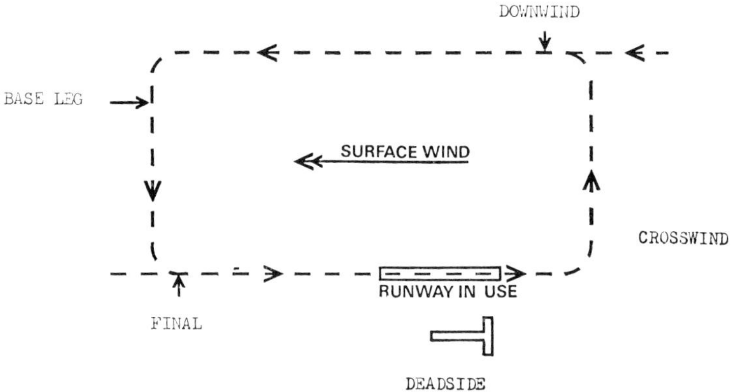


The AFIS Officer will answer each report by either the aircraft call-sign or 'Roger' and pass any traffic information.

| *ACFT* | **'G-BO downwind'** |
|---|---|
| *AFIS* | 'Roger'<br><br>'G-BO one Cessna 172 turning base, one Cessna 150 taking-off' |

6.4 **Arriving aircraft**

(a) *Approximately 5 minutes from destination aerodrome or when handed over by ATCC*

The pilot should pass his details and route see Chapter 5 para. 4.4.

| | |
|---|---|
| *ACFT* | **'Wick this is GBFBO for joining instructions** |
| *AFIS* | 'GBFBO Wick Information pass your details and route' |

| | |
|---|---|
| *ACFT* | **'GBFBO Cessna 310 Abeam Duncansby Head at 1500 feet QNH 1009 VMC from Sumburgh to Wick estimating at 23-180 knots'** |

| | |
|---|---|
| *AFIS* | 'GBFBO. Runway 09. Circuit direction left hand surface wind 100 degrees 10 knots QNH 1014 QFE 1010 over' (plus traffic information) |
| *ACFT* | **'QNH 1014 QFE 1010 GBFBO'** |

(b) *Aerodrome in sight*

| | |
|---|---|
| *ACFT* | **'G-BO Aerodrome in sight joining downwind Left Hand Runway 09'** |
| *AFIS* | 'G-BO. Surface Wind 100 degrees 15 knots' (plus any traffic information) |

(c) *In circuit*

| | |
|---|---|
| *ACFT* | **'G-BO Downwind'** |
| *AFIS* | 'Roger'*<br><br>*or*<br><br>'Report Base Leg' |
| *ACFT* | **'G-BO Base Leg'** |
| *AFIS* | 'Roger' |
| *ACFT* | **'G-BO Finals'** |
| *AFIS* | 'Roger 270 Degrees 10 Knots'* |

*Traffic information passed if necessary.

(d) *After landing*

| | |
|---|---|
| *ACFT* | **'G-BO is Clear of Runway'** |
| *AFIS* | 'G-BO' (plus any essential aerodrome information) |

6.5 *Pilot wishes to transit the Aerodrome Traffic Zone (ATZ)*

The pilot should pass details see Chapter 5 para 4.4.

| | |
|---|---|
| *ACFT* | **'Wick this is GBFBO for Transit'** |
| *AFIS* | 'GBFBO Wick Information pass your Details' |

| | |
|---|---|
| *ACFT* | **GBFBO Cessna 310 Passing Abeam Duncansby Head at 1000 Feet Cromarty 1009 VMC** |
| *AFIS* | 'GBFBO may transit at 1000 Feet Report overhead the Airfield'<br>*or*<br>'GBFBO negative. Remain outside the ATZ due to Ambulance Flight departing for Kirkwall. Over' |

## 7 PENETRATION OF MILITARY AERODROME TRAFFIC ZONES

Details of Military Aerodrome Traffic Zones (MATZ) are to be found in the AIP (RAC) section.

15nm or 5 minutes flying time (whichever is the greater) from the zone boundary the pilot establishes communication with the controlling aerodrome and passes his details. See Chapter 5 para 4.4.

| | |
|---|---|
| *ACFT* | **'Brize Norton Approach this is GBFBO Request MATZ Penetration'** |

| | |
|---|---|
| *APP* | 'GBFBO Brize Norton Approach Pass Your Details' |

The pilot is then required to:

(a) Comply with any instructions issued by the controller.

(b) Maintain a listening watch on the allocated RTF frequency until the aircraft is clear of the MATZ.

(c) Advise the controller when the aircraft is clear of the MATZ.

The expression 'CLUTCH QFE' may be passed by ATC and indicates the QFE of the higher, or highest, aerodrome of a combined zone.

Outside the hours of service published in AIP (RAC) section pilots are advised to proceed with caution if no reply is received after two consecutive RTF calls to the ATSU.

## 8 AIREPS

An Air-Report (air rep) is a standard form of position report consisting of three sections as follows:

| | |
|---|---|
| Section 1 | *Position information*<br>Aircraft callsign<br>Position<br>Time<br>Flight level or altitude<br>Next position and time over |
| Section 2 | *Operational information*<br>Estimated time of arrival<br>Endurance |
| Section 3 | *Meteorological information*<br>Air Temperature<br>Wind<br>Turbulence<br>Aircraft icing<br>Supplementary information |

A complete example is:

| | |
|---|---|
| *ACFT* | **'AIREP Speedbird 561 Position 49 North 050 West at 1317. Flight Level 310. Next Position 50N 040 West at 55 Endurance 0830 Temperature Minus 47 Wind 255/65. Turbulence Moderate. Scattered Cumulonimbus Top Flight Level 280'** |

Certain items may be omitted dependent on the type of airspace and the requirement for Met information. Details are in the AIP (RAC and MET sections).

## 9 OPERATIONS NORMAL

When 'Operations Normal' reports are transmitted by aircraft they should consist of the prescribed call followed by the words 'Operations Normal'. Full procedures are in ICAO Doc 4444.

| | |
|---|---|
| *ACFT* | **'. . .Control This is GBFBO. Operations Normal'** |
| *ATSU* | 'G-BO' |

CHAPTER 6

# PHRASEOLOGY VFR

## 1 APPROACH/ZONE CONTROL

### 1.1 General

VFR flights will be passed information on pertinent known traffic to assist the pilots to maintain separation from both IFR and other VFR flights.

If other flights are letting down on an instrument approach aid VFR flights will not be given clearance for a straight in approach and will be advised to avoid the let-down area.

Since such information can take many different forms only one example is shown:

| | |
|---|---|
| *APP* | 'G-BO an Islander is leaving the Zone to the South East VFR and a B707 is on ILS final approach at 8 Miles—Keep South of the final approach track. Report Aerodrome in sight' |

### 1.2 Departing flights

These will be given information on other traffic and are to report leaving the area of jurisdiction of approach/zone control. At this point details of the local Altimeter Setting Region QNH and the FIS frequency will be passed.

| | |
|---|---|
| *ACFT* | **'G-BO Passing the Zone Boundary/ Leaving your Area/Frequency'** |
| *APP* | 'G-BO Chatham QNH 1010 Flight Information available with London Information 124.6' |
| *ACFT* | **'QNH 1010 and 124.6'** |

## 1.3 Arriving Flights

(a) *Establishment of Communication*

Arriving flights should normally establish communication with Approach/Zone Control when at least 15nm or 5 minutes flying time from the aerodrome traffic zone boundary. If unable to communicate with Approach/Zone it is acceptable for the pilot to call the TWR to pass details of the flight.

If radar sequencing of IFR flights is in progress ATC will provide VFR flights with information to enable them to fit into the traffic sequence.

(b) *Instructions to join the Approach area or Zone*

Communication is established and details and route passed. See Chap 5 para 4.4.

| | |
|---|---|
| *ACFT* | **'Stansted this is GBFBO for Joining Instructions'** |
| *APP* | 'GBFBO Pass Your Details and Route' |
| *ACFT* | **'GBFBO Cessna 172 over Cambridge heading 210 2000 ft Chatham 1010 VFR from Newmarket to Stansted estimating at 35 request straight in approach on runway 23 120 knots'** |
| *APP* | 'G-BO cleared VFR to Stansted for straight in approach Runway 23 Surface Wind 270 degrees 10 knots QFE 1010 Report Aerodrome in sight' |
| *ACFT* | **'Runway 23 QFE 1010 G-BO'** |

Instructions to Special VFR flights may also include a maximum altitude/height to fly e.g. 'NOT ABOVE 2000 FEET QNH 1010'. The next report to Approach would be the destination sighting report and the Approach Controller would then normally transfer the aircraft to the Tower for landing instructions.

| | |
|---|---|
| *ACFT* | **'G-BO Aerodrome in sight'** |
| *APP* | 'G-BO Contact Tower 118.4' |
| *ACFT* | **'Tower 118.4 G-BO'** |

(c) *Landing or circuit joining instructions*
These normally take the form of a clearance to position for a straight in approach or to join the aerodrome traffic circuit at a specific point (see Chapter 5 para 3.1). Information on other aircraft in the circuit will be passed.

| | |
|---|---|
| *ACFT* | **' . . . Tower GBFBO Aerodrome in sight Request Landing Instructions'** |
| *TWR* | 'G-BO Clear for Straight In Approach Runway 23 QFE 1010 Report Long Final/Final' |
| *ACFT* | **'Wilco Runway 23 QFE 1010 G-BO'** |

| | |
|---|---|
| *TWR* | 'G-BO Join Downwind Left Hand Circuit for Runway 26. QFE/QFE Threshold 1010 Report Downwind' |
| *ACFT* | **'Wilco Downwind Left Hand. Runway 26 QFE 1010. G-BO'** |

| | |
|---|---|
| *TWR* | 'G-BO Join Left Base. Runway 26 QFE/QFE Threshold 1010 Number 2 to a Beagle Pup turning final report left base' |
| *ACFT* | **'Wilco Left Base Runway 26 QFE 1010—Pup in Sight. G-BO'** |

The controller will also pass information on work in progress or obstructions on or near the landing area.

Subsequent instructions to the aircraft in the circuit would be passed as shown in Chapter 5 para 3.1.

1.4 **Transit flights—VFR**

(a) *Establishment of Communication*
See para 1.3 (a)

(b) *Instructions to transit the Approach area or Zone*
Details are to be passed as in Chapter 5 para 4.4

| | |
|---|---|
| *ACFT* | **'Stansted Approach this is GBFBO to Transit Zone'** |
| *APP* | 'GBFBO Stansted Approach Pass your Details and Boundary Estimate' |
| *ACFT* | **'GBFBO Cessna 310 5 Miles South of Bedford heading 130 2000 ft Chatham 1010 VMC Zone Boundary at 10'** |

| | |
|---|---|
| *APP* | 'G-BO Clear to Cross VFR.<br>An Islander has just departed Runway 23 climbing to 3000 feet an route Birmingham Report passing the Zone Boundary' |
| *ACFT* | **'Cross VFR G-BO'** |

Pilots wishing to cross Control Zones in IMC are to request Special VFR clearance. A typical Special VFR clearance is:

| | |
|---|---|
| *APP* | 'G-BO is cleared to cross the Zone Special VFR not above 2000 feet QNH 1014' |

Some locations have special routes or visual reporting points and these would be included in the clearance where necessary.

CHAPTER 7

# PHRASEOLOGY—IFR

## 1 INTRODUCTION

Many units provide a radar control or radar advisory service to flights. Chapter 8 (RADAR) should be read in conjunction with this chapter.

## 2 CANCELLATION OF AN IFR FLIGHT PLAN

| | |
|---|---|
| *ACFT* | **'. . . Tower GBFBO. Cancel IFR Flight Plan'** |
| *TWR* | 'GBFBO. IFR Flight Plan Cancelled at 0952' |

If IMC is likely to be encountered along the route of the aircraft.

| | |
|---|---|
| *TWR* | 'GBFBO. IMC reported (or forecast) in the vicinity of . . .'(location) |

## 3 AERODROME CONTROL

### 3.1 Departure Approval

Where a departure needs to be regulated, one of the following phrases may be included in the clearance.

(a) 'CLEARANCE EXPIRES AT . . . ' (time)

This indicates that if the aircraft is not airborne by the time stated, a new clearance will have to be obtained.

(b) 'TAKE OFF NOT BEFORE . . . ' (time).

The pilot will be informed so that he can calculate the best time to start engines.

(c) 'UNABLE TO CLEAR AT . . . ' (Planned level)

Given when ATC is unable to clear the flight at the planned level. An alternative will be offered whenever possible.

(d) 'JOIN AIRWAYS AT . . . ' (place and level)

'NOT BEFORE . . . ' (time)

Used when an Airways clearance is given to an aircraft departing from an aerodrome outside controlled airspace.

3.2 **Standard Instrument Departures**

The AIP (RAC section) details those areodromes where Standard Instrument Departure (SID) procedures are in force.

| | |
|---|---|
| *TWR* | 'GBFBO. Cleared to Start<br>Cleared to Amsterdam<br>Clacton Two Zero Departure<br>Squawk Alpha 5511' |
| *ACFT* | **'Cleared to Amsterdam<br>Clacton Two Zero Departure<br>Squawk 5511. GBFBO'** |

'CLACTON TWO ZERO' is the SID.

At some airports using SIDs, aircraft may be transferred by the TWR direct to the ATCC and not to Approach.

3.3 **Reduced separation in the vicinity of an aerodrome**

This procedure is detailed in ICAO Doc 4444 part IV. The RTF involved requires pilots only to state that they can see the aircraft concerned and can maintain their own separation, or for the controller to give directions to effect separation.

## 4 APPROACH/ZONE CONTROL

### 4.1 Departing Flights

(a) *Transfer to FIS*

| | |
|---|---|
| *ACFT* | **'G-BO Passing the Zone Boundary'** |
| *APP* | 'G-BO.Chatham QNH 1010.* Flight<br>Information available with London<br>Information on 124.6' |

*If necessary, traffic information is passed.

(b) *Transfer to Airways*

On reaching the level or reporting point required by Approach Control, flights joining Airways will be transferred to the appropriate ATCC controller.

| | |
|---|---|
| *ACFT* | **'G-BO reaching flight level 80'** |
| *APP* | 'G-BO Contact London Control 125.8' |
| *ACFT* | **'London 125.8 G-BO'** |

4.2 **Arriving Flights**

(a) *Establishment of Communication*
The pilot of an arriving flight should establish communication with Approach Control when instructed to do so by the ATCC. If the aircraft is outside controlled airspace then communication should be established at least 10 minutes before the ETA at the aerodrome.

(b) *Instructions to join the Approach Area or Zone*
Aircraft flying within controlled airspace receive detailed instructions from the ATCC before being transferred. After establishing communication with Approach, the pilot will be given an expected approach time (where applicable), and descent and approach instructions as in (c) and (d)
Aircraft flying outside controlled airspace should establish communication with Approach Control and pass details and route see Chapter 5 para 4.4.

| | |
|---|---|
| *ACFT* | **'Birmingham Approach this is GBFBO for Zone Joining Instructions'** |
| *APP* | 'GBFBO Birmingham Approach pass your Details and Route' |
| *ACFT* | **'GBFBO Cessna 310 over Cheltenham Heading 030 flight level 50 IMC Exeter to Birmingham estimating zone boundary at 02 GM at 05 180 knots for ILS training'** |
| *APP* | 'G-BO remain outside controlled airspace time check 50' |

The joining instructions passed by Approach will be:
Aircraft callsign
Clearance Limit (normally the holding or approach facility)
Route (which may specify an entry point at the zone boundary, if any)
Flight Level or altitude (which may not be the cruising level of the aircraft)
Expected Approach Time (or an Onward Clearance Time from a holding point/Delay not determined/No delay expected)

| | |
|---|---|
| *APP* | 'GBFBO is cleared to the XC (Approach Facility)<br>From 5 miles south of XS (Entry Position)<br>via the XS (Route)<br>At flight level 50<br>Expected Approach Time 32<br>Report passing the Zone Boundary' |

The pilot must read back the full clearance.

(c) *Information to be passed to aircraft*

As soon as possible after establishment of communication the controller will pass the following:

| | |
|---|---|
| *APP* | 'G-BO this will be a Beacon (ILS) Approach<br>Runway 26<br>The 0920 Weather<br>220 degrees 10 knots<br>Visibility 4 kilometres<br>In Rain<br>6 OKTAS at 800 feet<br>QNH 1014 QFE 1013' |
| *ACFT* | **'QNH 1014 QFE 1013 G-BO'** |

See Chapter 9 for details of the Automatic Terminal Information Service.

(d) *Clearance to commence Instrument Approach*

(i) In the case of airports with non-mandatory instrument approach procedures the aircraft may request a straight in approach. Otherwise, the aircraft will complete the procedure as laid down in the AIP (RAC section).

| | |
|---|---|
| *ACFT* | **'G-Bo Request Straight in Approach on ILS'** |
| *APP* | 'G-BO Clear for Straight in Approach report established ILS'<br>*or*<br>'G-BO due to Traffic Maintain Flight Level 50' |

(ii) When airports have mandatory instrument approach procedures or when the aircraft intends to fly to the beacon and complete the laid down procedures:

| | |
|---|---|
| *APP* | 'G-BO Cleared for Beacon (ILS) Approach report Beacon outbound (Outer Marker Outbound)' |
| *ACFT* | **'Wilco G-BO'** |
| *ACFT* | **'G-BO Beacon (Outer Marker) Outbound at . . . ' (time)** |
| *APP* | 'Report Base Turn (Procedure Turn) Complete'<br>*or*<br>'Report Beacon (Outer Marker) Inbound' |
| *ACFT* | **'G-BO Beacon Inbound (Outer Marker Inbound)'** |
| *APP* | 'G-BO QFE 1010 Report Visual' |
| *ACFT* | **'QFE 1010 G-BO'** |
| *ACFT* | **'G-BO Visual'** |
| *APP* | 'G-BO Contact Tower 118.4' |
| *ACFT* | **'Tower 118.4 G-BO'** |

Note: Items in parenthesis are for ILS approach

Landing clearance will be passed to the pilot during the approach.

(e) *QGH Procedure*

(i) Initial Approach

| ITEM | | PHRASEOLOGY |
|---|---|---|
| Essential Information | *ACFT* | **'. . . Approach GBFBO Request QGH Letdown'** |
| | *APP* | 'G-BO Transmit for DF' |
| | *ACFT* | **'G-BO Transmitting for QGH Over'** |
| | *APP* | 'G-BO Make Your Heading . . . Degrees. Report Steady' |
| | *ACFT* | **'G-BO Turning To . . . ' (degrees) 'G-BO Steady on . . . ' (degrees)** |
| | *APP* | 'G-BO This will be a QGH Procedure Safety Altitude is . . . feet Aerodrome QNH is . . . millibars Runway in use is . . . Obstacle Clearance Limit/Visual Manoeuvring Height is . . . feet Check Your Decision Height' |
| | *ACFT* | **'QNH . . . Millibars G-BO'** |
| Weather Information | *APP* | 'The 0920 Weather. Surface Wind 260 degrees 10 knots 6 kilometres in Rain 5 Oktas 1000 feet' |
| | *ACFT* | **'G-BO'** |
| Homing | *APP* | 'G—BO Transmit for DF' |
| | *ACFT* | **'One Two Three G-BO'** |
| | *APP* | 'G-BO Turn Left/Right Heading . . . ' (degrees) |
| | *ACFT* | **'Turning Left/Right Heading . . . ' (degrees) G-BO'** |

The homing procedure will then be repeated as necessary until the VDF indication shows that the aircraft is overhead.

(ii) Intermediate Approach

| ITEM | | PHRASEOLOGY |
|---|---|---|
| Overhead turn. | *APP* | 'G-BO DF Indicates you have passed Overhead turn Left/Right Heading . . . (degrees) Report Steady' |
| | *ACFT* | **'Turning on to . . . (degrees) G-BO'**<br>**'Steady Heading . . . (degrees) G-BO'** |
| | *APP* | 'G-BO QFE (QFE Threshold) . . . (mb) Acknowledge' |
| | *ACFT* | **'QFE . . . (mb) G-BO** |
| Outbound descent | *APP* | 'G-BO descend to . . . feet on QFE Report Level' |
| | *ACFT* | **' . . . feet G-BO'** |
| | *APP* | 'Transmit for DF'<br>'Turn Left/Right Heading . . . ' (degress) } *repeated as necessary* |
| Inbound turn | *APP* | 'G-BO Prepare for Inbound Turn Turn Left/Right Heading . . . (degrees) Report Steady' |
| | *ACFT* | **'Turning to . . . (degrees) G-BO'**<br>**'Steady . . . (degrees) G-BO'** |

(iii) Final Approach

| ITEM | | PHRASEOLOGY |
|---|---|---|
| Inbound descent | *APP* | 'G-BO in the Event of Missed Approach . . . ' (appropriate instructions) |
| | *ACFT* | **' . . . (repeat back) G-BO'** |
| | *APP* | 'G-BO Descend to Decision Height Report Aerodrome in Sight' |
| | *ACFT* | **'Wilco'** |

(iii) Final Approach (cont.)

| ITEM | | PHRASEOLOGY |
|---|---|---|
| inbound descent | *APP* | 'Transmit for DF'<br>'Turn Left/Right Heading . . . ' (degrees) } *repeated as necessary*<br>*or*<br>'Continue heading . . . degrees'<br><br>Plus landing or circuit instructions as required. |
| | *ACFT* | **'G-BO Aerodrome in Sight'** |
| | *APP* | 'Contact tower 118.4' |
| | *ACFT* | **'Tower 118.4'** |

(iv) Missed Approach

If the VDF indicates to the controller that the aircraft has passed overhead but it has NOT reported visual:

| | |
|---|---|
| *APP* | 'G-BO Overshoot, I Say Again, Overshoot . . . (appropriate instructions) Acknowledge' |
| *ACFT* | **'G-BO Overshooting'** |

**(f)** VDF Procedure

This pilot interpreted procedure is detailed in the AIP (RAC section)

4.3 **Transit Flights**

(a) *Establishing of Communication.*
As in para. 4.2 (a)

(b) *Instructions to cross the Approach Area or Zone*

Pilots should pass both details and route information, see Chapter 5 para. 4.4, and include an ETA for the zone boundary (if applicable) and the approach or en-route facility.

| | |
|---|---|
| *ACFT* | **'. . . Zone, GBFBO to Transit Zone'** |
| *APP* | '. . . Zone, GBFBO Pass your details and route' |
| *ACFT* | **'GBFBO Cessna 310 over Portsmouth Heading 250 flight level 45 IMC from Shoreham to Exeter zone boundary at 55 170 Knots'** |
| *APP* | 'G-BO From . . . (entry position)<br>to . . . (exit position) Via<br>Fawley . . . (facility) to Maintain Flight Level 50 Report Passing the Zone Boundary' |

The pilot must give a full read back.

## 5 AIRWAYS

### 5.1 General

Most phraseology used on airways has been covered in Chapter 5—General Phraseology. Certain ATC instructions are applicable to Airways (or oceanic) Flights. 'LOSE TIME TO ARRIVE OVER . . . (reporting point) at . . . (time) 'ARRANGE YOUR FLIGHT TO ARRIVE OVER . . . (reporting point) at . . . (time) 'CROSS . . . (reporting point) NOT LATER THAN . . . (time)'

### 5.2 En-Route Holding

Pilots are required to report in the following way:

(a) The time and level of reaching a specific holding point to which cleared:

| | |
|---|---|
| *ACFT* | **'London Control G-BO Daventry at 05 flight level 110'** |

(b) When leaving a holding point:

| | |
|---|---|
| *ACFT* | **'London Control G-BO Leaving Daventry at 15'** |

(c) When vacating a previously assigned level for a new assigned level:

| | |
|---|---|
| *ACFT* | **'London Control G-BO leaving flight level 110 for flight level 90'** |

5.3 **Flights Crossing Airways**

Detailed procedures are contained in the AIP (RAC section).

**Initial Contact**

(a) Aircraft callsign and request for crossing clearance
(b) Crossing position

| | |
|---|---|
| *ACFT* | **'London Information this is GBFBO for crossing clearance at Daventry (position)** |

On request from the ATSU the pilot will pass his details (see Chapter 5 para. 4.4) the requested flight level and ETA for the crossing position.

| | |
|---|---|
| *ATSU* | 'GBFBO London Information pass your message' |
| *ACFT* | **'GBFBO Cessna 310. 36 miles north-east Daventry heading 240 flight level 120 IMC to cross Daventry at flight level 100 at 1305'** |

Unless otherwise requested by ATC the aircraft crossing Airways will remain in communication with the FIS controller and, after obtaining clearance, will report when the aircraft is estimated to be at the nearest boundary of the airway:

| | |
|---|---|
| *ACFT* | **'London G-BO crossing amber one position 5 miles east of Daventry at 03. Flight level 100'** |
| *ATSU* | 'G-BO report leaving the airway' |

5.4 **Flights joining Airways**

**Initial contact**

(a) Aircraft callsign and request for joining clearance
(b) Entry position

| | |
|---|---|
| *ACFT* | **'London control this is GBFBO for joining clearance at Daventry'** |

The aircraft will be asked to pass his details and route, see Chapter 5 para 4.4.

| | |
|---|---|
| *ACFT* | **'G-BO Cessna 310 20 miles north-east Daventry heading 240 flight level 120 IMC from Tollerton Daventry at 45 amber one east amber one to Shoreham 180 knots request flight level 130'** |

5.5 **Flights leaving Airways**

(a) *Outside controlled airspace*

| | |
|---|---|
| *ATCC* | 'G-BO is cleared to leave controlled airspace 5 miles west of Ibsley to maintain flight level 120 report Ibsley' |

(b) *On transfer to an Approach Control Unit*

| | |
|---|---|
| *ATCC* | 'G-BO is cleared to the . . . (approach or holding facility)<br>Via . . . (route)<br>To maintain flight level . . . (or descent instructions)<br>Expected approach time . . . /*or* no delay expected<br>Contact . . . (approach unit) at . . . (position/time/level) |

EAT's may not be given by the Area unit if the overall delay is 20 minutes or less.

## 6 OCEANIC (TRANSATLANTIC) PHRASEOLOGY AND PROCEDURES

### 6.1 General

Full details of procedures to be followed by aircraft on Transatlantic crossings are detailed in AIP (RAC section 8)

### 6.2 Clearances—in flight requests

Requests for clearance should include:
Aircraft callsign
Oceanic Control Area (OCA) entry point and ETA
Requested flight level and Mach Number
Any change to flight plan affecting OCA

| | |
|---|---|
| *ACFT* | **'Shanwick Oceanic this is Speedbird 561 from London**<br>**Estimating 57 north 10 west at 1021**<br>**Requesting flight level 350**<br>**Mach 0.84**<br>**via track Bravo'** |

(a) *En-Route*

Control information is relayed through a communication centre. Messages will always indicate the originator of the message.

| |
|---|
| 'Speedbird 561 Shanwick clears you to . . . ' (clearance) |
| 'Speedbird 561 Shanwick clears you to climb/descend to . . . advise leaving/reaching' |
| 'Speedbird 561 Shanwick relay for Gander . . . ' (message) |

6.3 **Position Reporting**

| | |
|---|---|
| *ACFT* | **'Shanwick Oceanic this is Speedbird 561<br>Position 57 north 10 west at 1021<br>Flight level 350<br>Estimating 58 north 20 west at 1101'** |

See Chapter 5 para 8 for full AIREP procedures.

6.4 **Emergency Frequency Watch**

The Emergency Frequency of 121.5 MHz should be continuously guarded except for those periods when aircraft are carrying out communications on other channels or when airborne equipment or cockpit duties do not permit the simultaneous guarding of two channels.

6.5 **Supersonic transport flights** (See para 8)

(a) *Position reporting:*
Full position reports are required at the Oceanic entry and exit points and at 30W and 50W (as in Chapter 5 para 4.2)
(b) *Abbreviated position reports* are expected at 20W, 40W and 60W and consist of:

Aircraft callsign
Position (which may be reported by track code letter and longitude)
Time

| *ACFT* | **'Speedbird 171 SM 30 at 1440'** |
|---|---|

7 **SELECTIVE CALLING (SELCAL) PROCEDURES**

7.1 **General**

Voice calling is replaced by the transmission of coded tones over the RTF channel in use. A single selective call lasts for 2 seconds and consists of four pre-selected audio tones generated by the aeronautical station coder and decoded in the airborne receiver. Receipt of the assigned SELCAL code activates a cockpit calling system of light and/or chime signals. SELCAL is used for ground to air selective calling. Aircraft use voice calling in every case. The pilot is still able to maintain a conventional listening watch if required.

7.2 **Notification of SELCAL codes**

It is the responsibility of the aircraft commander to ensure that all the ground stations with which it would normally communicate are advised of the SELCAL code associated with its RTF callsign.

7.3 **Pre flight check**

The aircraft should contact the aeronautical station and request a pre-flight SELCAL check and, if necessary, should advise the aeronautical station at that time of its SELCAL code. Where appropriate the aeronautical station will advise the aircraft of the primary and secondary frequencies. A SELCAL check will be made on the SECONDARY frequency first and PRIMARY frequency next. After the check the aircraft is ready for communication on the primary frequency. It may not always be possible to initiate SELCAL check by a call on VHF, in which case the call may be made direct on the HF channel.

| | |
|---|---|
| *ACFT* | **'Shanwick radio this is Speedbird 579 check SELCAL on 8854'** |
| *A/G Station* | 'Speedbird 579 this is Shanwick radio checking SELCAL on 8854' (transmits appropriate SELCAL code for BA 579) |
| *ACFT* | **'Speedbird 579 SELCAL OK— request SELCAL check on 5638' (Procedure repeated)** |

### 7.4 En route procedures

Aircraft initiate calls by normal voice calling, some aeronautical stations use SELCAL to initiate calls for scheduled reports.

Once a SELCAL guard has been assured with a particular aircraft, SELCAL will normally be employed by aeronautical stations whenever they need to call the particular aircraft concerned.

If the SELCAL signal remains unanswered by the aircraft after 2 calls on the primary frequency and 2 calls on the secondary the aeronautical station should revert to voice calling. If both SELCAL and voice calling prove unsuccessful the station should ask other stations to call the aircraft.

The aircraft shall ensure that the aeronautical stations concerned with its flight are immediately made aware of any malfunctioning of its SELCAL installation and that voice calling is necessary.

All stations are to be advised when the SELCAL installation is again functioning normally.

## 8 RTF PHRASEOLOGY FOR SUPERSONIC TRANSPORT AIRCRAFT

### 8.1 Initial call on transfer to an ATS unit while in supersonic flight

| | |
|---|---|
| *ACFT* | **'. . . (acft callsign) Supersonic . . .' (flight level)** |

Information on 'position and time over' may be required

## 8.2 Commencement of transonic acceleration and climb

| | |
|---|---|
| *ACFT* | **'Request climb at . . . (time or position)<br>to Supersonic . . .' (flight level)** |
| *ATSU* | 'Cleared climb at . . . (time or position) to Supersonic . . . (flight level)<br>Report starting acceleration' |
| *ACFT* | **'Starting acceleration at . . .' (time or position)** |

## 8.3 Cruise climb

| | |
|---|---|
| *ACFT* | **'Request cruise climb between . . . (flight level) and . . .' (flight level)** |
| *ATSU* | *'Cleared cruise climb between . . . (flight* level) and . . .' (flight level) |

| | |
|---|---|
| *ACFT* | **'Request cruise climb above . . .' (flight level)** |
| *ATSU* | 'Cleared cruise climb above . . .' (flight level) |

CHAPTER 8

# PHRASEOLOGY—RADAR

## 1 INTRODUCTION

Radar services are provided after an aircraft has been identified. The phrase 'Identified' may be omitted by area radar units where special arrangements exist.

Any turns given as part of a radar service will be 'right' or 'left'. Radar headings must be read back unless otherwise instructed by the radar controller, eg when carrying out a radar approach.

## 2 POSITION INFORMATION

2.1 Pilots will be informed of their position:

(a) On first identification (except when 8.1 applies).

(b) When the pilot requests.

(c) When the aircraft is flying off the correct track.

(d) When an aircraft estimate differs significantly from the radar controllers estimate based on radar observation.

(e) When the pilot is instructed to resume his own navigation following radar vectoring.

(f) When on approach to land at least once during each leg of the circuit and, if on straight in approach, once before commencement of descent on final.

2.2 The position will be in one of the following forms

(a) A well known geographical position.

(b) Bearing (using points of the compass) and distance from a known position.

(c) Magnetic track and distance to a reporting point, an en-route navigational aid, or approach aid.

(d) Latitude and longitude (by certain area radar units).

(e) Distance from touchdown if aircraft is on final approach.

| | |
|---|---|
| *RADAR* | (a) 'Over Daventry'<br>(b) '10 miles north of Stansted'<br>(c) 'Magnetic track to Daventry is 130 range 20 miles'<br>(d) '52:30 north 01:40 west'<br>(e) '7 miles from touchdown' |

## 3 RADAR IDENTIFICATION

| *Circumstances* | Controller Phraseology |
|---|---|
| When the position of the aircraft is unknown<br><br>When a turn for identification is required<br><br>If the aircraft is in an orbit | 'What is your heading and level'<br><br>'For identification turn left/right heading . . . (degrees) Report steady'<br>Note. a period of time eg 'For one minute' may be stipulated<br><br>'What heading are you passing and direction of turn'<br>*then*<br>'For identification stop the turn heading . . . (degrees) report steady' |
| *To obtain a VDF bearing* | 'Transmit for DF' |
| On identification<br>1. After turn<br><br>2. Immediately after take-off<br>3. Using aircraft position report | <br>'Identified . . . miles N/E/S/W of . . . ' (position)<br><br>'Identified . . . ' (further instructions as required) |
| If aircraft is not identified after a turn | 'Not identified. For identification turn left/right heading . . . ' (degrees)<br>*or*<br>'Not identified. Resume own navigation' |
| If no aircraft is seen | 'No radar contact . . . I will keep you advised . . .'<br>(instructions or advice) |

| *Circumstances* | Controller Phraseology |
|---|---|
| To inform the pilot that Radar Control Service is being given (ie radar separation inside controlled or special rules airspace) | 'Under radar control'<br><br>*or*<br>'This is radar control service' |
| To inform the pilot that Radar Control Service has been terminated (ie non-radar separation is being applied) | 'Radar control service terminated. contact . . . (non-radar unit) on . . . ' (frequency) |
| To inform the pilot that Radar Advisory Service is being applied | 'Under radar advisory service'<br>*or*<br>'This is radar advisory service' |
| To inform the pilot that radar services are terminated | 'Radar Service terminated . . . ' (reason) |
| When vectoring is completed | 'Resume own navigation. position . . . ' (*or* 'Magnetic track to . . . (reporting point) range . . . miles'<br>This transmission may be prefaced by the phrase 'Radar Control/ Service terminated' if necessary |
| Prior to a radar handover to another radar unit | '. . . miles N/E/S/W of . . . (destination or reporting point) continue heading . . . and contact . . . radar on . . . ' (frequency) |
| Prior to passing out of radar cover | 'Will shortly lose radar contact. Contact . . . (ATSU) on . . .' (frequency). (A position would normally be passed) |

8.4 (cont.)

| *Circumstances* | Controller Phraseology |
|---|---|
| When speed adjustments are necessary | 'Increase speed to . . . ' (knots)<br>*or*<br>'Reduce speed by/to . . . ' (knots)<br>*or*<br>'Maintain present speed (to . . .)'<br>*or*<br>'Resume normal speed'<br>*or*<br>'Reduce to minimum approach speed' |
| Azimuth instructions | 'Turn left/right heading . . . ' (degrees)<br>*or*<br>'Continue the left/right turn heading . . . ' (degrees)<br>*or*<br>'Turn left/right . . . (number of degrees) heading . . . ' (degrees) |
| Delaying action | 'Make a 360 degree turn to the left/right . . . ' (reason)<br>*or*<br>'Orbit to the left/right . . . ' (reason) |
| Navigational assistance to aircraft lost or off track | 'Your position is . . . '<br>*or*<br>'Magnetic track to . . . (position) is . . . (degrees) distance . . . miles' |
| To give a gyro check if significant error is suspected | 'Check your gyro heading. Magnetic track made good appears to be . . . ' (degrees) |

## 5 TRAFFIC INFORMATION AND AVOIDANCE

5.1 Information shall include the following:

(a) Bearing from the aircraft in terms of the 12 hour clock (when the aircraft is in a turn the position of the other aircraft by compass points).

(b) Distance from the aircraft in miles.
(c) Direction in which the other aircraft is proceeding.

| *RADAR* | 'G-BO traffic at 2 o'clock 5 miles crossing right to left'<br>*or*<br>'G-BO traffic east of you at 5 miles crossing left to right' |
|---|---|

5.2 **Traffic avoidance**

| *Circumstance* | Radar Phraseology |
|---|---|
| Avoiding action | 'G-BO avoiding action. Turn left/right immediately heading . . . (degrees) . . .' (reason) |
| To indicate to a pilot the necessary avoiding action if he does not have the other traffic in sight | 'G-BO unknown traffic . . . o'clock at . . . miles . . . (head-on/crossing left/right). If not sighted turn left/right heading . . .' (degrees)—(if appropriate) |
| When all collision risk has passed | 'G-BO now clear of traffic. Turn left/right heading . . .' (degrees)<br>*or*<br>'G-BO now clear of traffic resume own navigation' (a position check will be passed) |

## 6 WEATHER REPORTING AND AVOIDANCE

Controllers offer advice and instructions concerning weather observed on radar or reported by aircrews as follows:

(a) Notify the pilot if radar indicates that there is weather ahead of the aircraft.

(b) Offer and supply circumnavigational assistance.

(c) Advise the pilot if routeing around weather will take the aircraft outside controlled airspace. The pilot will then decide whether to accept re-routeing.

(d) Provide navigational assistance if necessary.

(e) Provide radar advisory service if required.

| | |
|---|---|
| *RADAR* | 'G-BO radar indicates weather 20 miles ahead of you. A 10 degree turn to the left will take you to the west of the weather.' |
| *ACFT* | **'G-BO is well above cloud and can continue on track.'** |

## 7 LOSS OF COMMUNICATION OR RADAR CONTACT

Detailed procedures are in the AIP (RAC section).

### 7.1 Aircraft radio failure

| | |
|---|---|
| RADAR | 'G-BO reply not received. If you read . . . (radar unit callsign) turn left/right heading . . . I say again turn left/right heading . . . ' |

| | |
|---|---|
| RADAR<br>*(If SSR Available)* | 'G-BO reply not received. If you read . . . (radar unit callsign) SQUAWK A7600'<br>*or*<br>'G—BO if you read . . . SQUAWK ident' |

If, after this transmission has been made the aircraft is seen to follow instructions:

| | |
|---|---|
| RADAR | 'G-BO turn observed. I will continue to pass instructions.'<br>*or*<br>'SQUAWK observed I will continue to pass instructions' |

### 7.2 Loss of radar contact

| | |
|---|---|
| RADAR | 'I have lost radar contact due . . . (reason) I will advise you when contact regained . . . ' (alternative instructions)<br>*or*<br>'Radar contact will be broken for . . . (minutes) due . . . (reason). I will advise you when contact regained . . . ' (alternative instructions). |

## 8 PHRASEOLOGY EXCLUSIVE TO APPROACH RADAR CONTROL

### 8.1 Departure instructions

These may include radar headings and level restrictions:

| | |
|---|---|
| RADAR | 'G-BO turn right heading 360 climb to flight level 50' |

The clearance passed by TWR may be in the form:

| | |
|---|---|
| TWR | 'G-BO after departure turn right heading 360 to climb when instructed by radar to flight level 50' |

### 8.2 Special VFR and VFR flights

Certain aerodromes specify in the AIP (RAC section) that SVFR flights will be directed by radar when flying within the controlled airspace around those aerodromes.

Where radar sequencing of IFR flights is in operation the pilots of VFR flights will be provided with information to enable them to fit into the landing sequence.

### 8.3 Radar vectoring to Final Approach

*TYPICAL PATTERN*

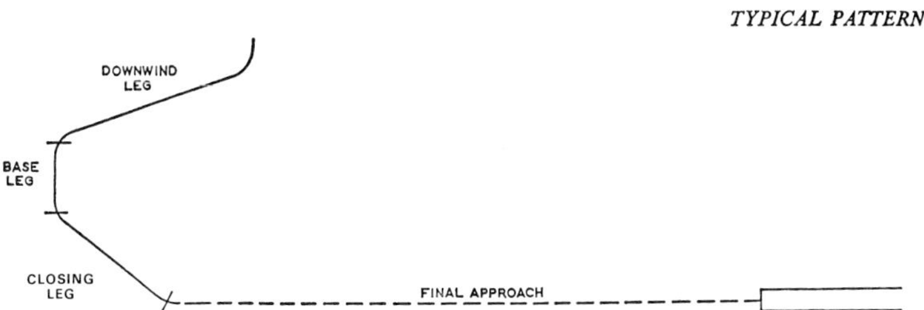

8.3 (cont.)

| *Circumstance* | Radar Controller Phraseology |
|---|---|
| When a radar approach is impracticable or not available | '. . . (type) approach not available due to . . . ' (reason—equipment failure, clutter etc.) |
| When a radar approach can probably be completed but there is a possibility of the aircraft response being lost in clutter: | |
| (a) When ILS is usable | '. . . (type) approach may be affected by clutter. Advise you monitor approach with ILS' |
| (b) When ILS is not usable | '. . . (type) approach may be affected by clutter. Overshoot instructions will be passed in good time if necessary' |
| Radar vectoring for a Surveillance radar approach | 'This will be radar vectoring for a surveillance radar approach runway . . . terminating at half/two miles from touchdown. Obstacle clearance limit . . . feet. Check your minima' |
| Radar vectoring for ILS approach | 'This will be radar vectoring to ILS runway . . . ' |
| Radar vectoring for ILS approach with glidepath inoperative. | 'This will be radar vectoring for a localiser only approach. Runway . . . ' |
| Before reaching Base Leg | (a) 'Descend to . . . feet' (Intermediate approach altitude) QNH . . . ' (millibars)<br><br>(b) 'This is a left/right hand circuit runway, . . . '<br><br>(c) 'On downwind leg . . . miles N/S/E/W of . . . ' (name of aerodrome) |

8.3 (cont.)

| *Circumstance* | Radar Controller Phraseology |
|---|---|
| On Base Leg | 'On left/right base . . . miles N/S/E/W of . . .' (name of aerodrome) |
| When radio failure procedure to be adopted is not published | 'If you lose radio contact on this approach . . . (alternative instructions) and contact . . . (ATSU) on . . .' (frequency) |
| In the case of a radar datum | 'Heights will be converted to QNH and altitude. Aerodrome/ threshold elevation is . . . feet' |
| On closing leg | 'Closing final approach track from the left/right . . . miles from touchdown'<br>the aircraft may also be descended to intercept the glidepath |
| Turn on to ILS localiser<br>(a) (Controller turn on) | 'Turn left/right heading . . . (degrees) report established on the localiser' |
| (b) (Pilot turn on) | 'Closing the localiser from the left/right. Report established' |
| Pilot reports established on the localiser | 'Continue descent on ILS. Range . . . miles from touchdown QFE . . . contact tower . . .' |
| Pilot reports established on the localiser. (Glidepath inoperative) | '. . . miles from touchdown, cleared for a localiser only approach, contact tower on . . .' |
| If glidepath inoperative | 'Advisory height and range information is available if required' (At radar controller's discretion) |
| Delaying action | 'Taking you through the localiser to establish from the left/right' |

Distances from touchdown will be passed at intervals.

## 8.4 Surveillance Radar Approach (SRA)

(a) *Introduction*
On a SRA the pilot is given accurate ranges from touchdown and advisory height (or altitude) information so that he can adjust his rate of descent as necessary.

(b) *Advisory Heights/Altitudes*
Advisory heights are given which assume a glidepath angle of approximately 3 degrees with the aircraft commencing descent from a height of 1500 feet. If commencing descent from other heights the descent instruction will be issued at the appropriate range (eg 2000 ft—6½nm, 1100 ft—3½nm). Advisory heights and ranges at which descent is commenced, will be adjusted for other glidepath angles.

The radar controller will assume that aircraft will be flying on QFE datum and this will be emphasised by the controller inserting the words 'QFE' after the advisory heights at intervals during the approach.

If a pilot advises the controller that he will be using QNH for the approach then the aerodrome or threshold elevation will be added to the advisory heights and the total rounded up to the next 25 feet. The word 'ALTITUDE' will be used instead of 'height' whenever applicable.

In the case of the SRA terminating at ½ mile from touchdown, no advisory heights/altitudes will be passed after the aircraft passes a range of 1 mile from touchdown.

| Range from Touchdown (nm) | Advisory Height (feet) | | |
|---|---|---|---|
| | 2¾° Glidepath | 3° Glidepath | 3¼° Glidepath |
| 4½ | 1350 | 1400 | 1575 |
| 4 | 1200 | 1250 | 1400 |
| 3½ | 1050 | 1100 | 1225 |
| 3 | 900 | 950 | 1050 |
| 2½ | 750 | 800 | 875 |
| 2 | 600 | 650 | 700 |
| 1½ | 450 | 500 | 525 |
| 1 | 300 | 350 | 350 |

(c) *Landing clearance*

The clearance to land will normally be passed to the pilot before the aircraft reaches 2 miles from touchdown.

(d) *To alter course of the aircraft*

| | |
|---|---|
| RADAR | 'Turn left/right heading . . .' (degrees)<br>'Turn left/right . . . (number of degrees) . . . heading 260' |

(e) *To indicate displacement of the aircraft about the final approach*

| | |
|---|---|
| RADAR | 'Heading is good'<br>'Heading 260 is good'<br>'Closing toward track from the left/right'<br>'On track'<br>'Left/right of track' |

(f) *Surveillance Radar Approach terminating at 2nm from touchdown*

In the following example the aircraft—callsign GBFBO—is at 1500 feet (QFE) and has already been vectored to intercept the final approach track of runway 26 at 8nm from touchdown.

| *Situation* | *Radar Phraseology* | *Pilot Phraseology* |
|---|---|---|
| *Turn on to final* | 'G-BO turn right heading 260 final approach' | **'Turning right 260 G-BO'** |
| *8 miles from touchdown* | 'G-BO range 8 miles (contact Radar 123.25)' | **'G-BO' or 'Radar 123.25 G-BO'** |
| *Track adjustment* | 'G-BO left of track turn right 5 degrees heading 265' | **'Turning right 5 degrees heading 265 G-BO'** |

| *Situation* | *Radar Phraseology* | *Pilot Phraseology* |
|---|---|---|
| *7 miles from touchdown* | 'G-BO range 7 miles. Report approach lights or runway in sight' | **'Wilco'** |
| *6 miles from touchdown* | 'G-BO range 6 miles. Check wheels down and locked' | **'Wilco'** |
| *Approaching 5 miles from touchdown* | 'G-BO approaching 5 miles begin descent to maintain a 3 degree glidepath QFE 1013' | **'Leaving 1500 feet QFE 1013 G-BO'** |
| *4 miles from touchdown* | 'G-BO range 4 miles height should be 1250 feet. Clear to land RWY 26. Surface wind 260 at 10 knots' | **'Clear to land G-BO'** |
| *3 miles from touchdown* | 'G-BO range 3 miles height should be 950 feet check your decision height' | **'Wilco'** |
| *2 miles from touchdown* | 'G-BO range 2 miles height should be 650 feet. I cannot assist you further. Continue the approach or overshoot at your discretion' | **'Runway in sight landing G-BO'** |
| *Pilot wishes to continue on a visual approach* | 'G-BO contact tower 118.4' | **'Tower 118.4 G-BO'** |

Azimuth instructions as in 4 and 8.3 will be passed so that the aircraft is kept on, or close to, the final approach track.

Should the pilot decide to overshoot he should say 'OVERSHOOTING' and wait for instructions from the radar controller.

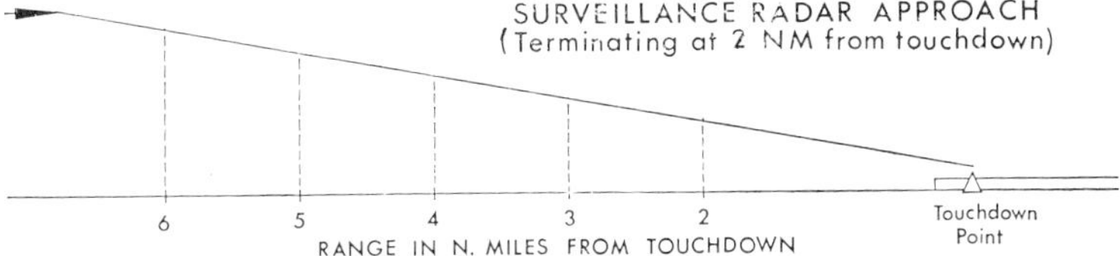


(g) *Surveillance Radar Approach terminating at ½nm from touchdown*

In the following example the aircraft—callsign GBFBO—is at 1500 feet (QFE) and has already been vectored to intercept the final approach track of runway 26 at 8nm from touchdown.

| *Situation* | *Radar Phraseology* | *Pilot Phraseology* |
|---|---|---|
| *Turn onto final* | 'G-BO Turn right heading 260 Final Approach' | **'Turning right 260 G-BO'** |
| *8 miles from touchdown* | 'G-BO range 8 miles (Contact Radar on 123.25)' | **'G-BO or Radar 123.25 G-BO'** |
| *Track Adjustment* | 'G-BO left of track turn right 5 degrees heading 265' | **'Turning right 5 degrees heading 265 G-BO'** |
| *7 miles from touchdown* | 'G-BO range 7 miles' | **'G-BO'** |
| *6 miles from touchdown* | 'G-BO range 6 miles check wheels down & locked' | **'Wilco'** |
| *Approaching 5 miles from touchdown* | 'G-BO approaching 5 miles begin descent to maintain a 3 degree glidepath QFE 1013' | **'Leaving 1500 feet QFE 1013 G-BO'** |

| *Situation* | *Radar Phraseology* | *Pilot Phraseology* |
|---|---|---|
| *4½ miles from touchdown* | 'G-BO range 4½ miles. Height should be 1400 feet' | **'G-BO'** |
| *4 miles from touchdown. Continuous transmission commenced* | 'G-BO range 4 miles. Height should be 1250 feet. Do not reply to further instructions clear to land RWY 26. 260 degrees 10 knots' | |
| *3½ miles from touchdown* | 'Range 3½ miles. Height should be 1100 feet' | |
| *3 miles from touchdown* | 'Range 3 miles. Height should be 950 feet' | |
| *2½ miles from touchdown* | 'Range 2½ miles. Height should be 800 feet' | |
| *2 miles from touchdown* | 'Range 2 miles. Height should be 650 feet. Check your decision height' | |
| *1½ miles from touchdown* | 'Range 1½ miles. Height should be 500 feet' | |
| *1 mile from touchdown*<br><br>*½ mile from touchdown* | 'Range 1 mile. Height should be 350 feet. On track Range ½ mile from touchdown. Approach completed. 260 degrees 10 knots Out' | |

Azimuth instructions will be issued to keep the aircraft on final approach track.

Should the pilot decide to overshoot he should say 'OVERSHOOTING' and wait for instructions from the radar controller.

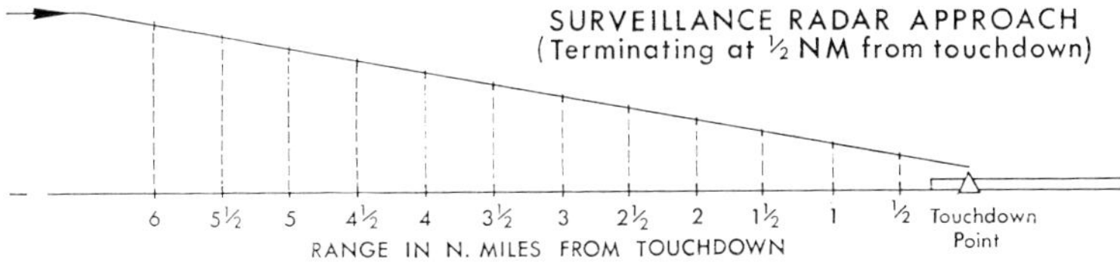


(h) *Missed approach*

Initiated by the controller if, during the latter stages of the radar approach the aircraft reaches a position from which it appears to the controller that a successful approach cannot be completed.

| | |
|---|---|
| *RADAR* | 'If unable to proceed visually— overshoot climb on heading . . . to . . . feet. . . .' (further instructions)<br><br>*or*<br><br>'Overshoot. I say again, Overshoot. Climb on heading . . . to . . . feet . . . (further instructions) Acknowledge'<br><br>*or*<br><br>'Climb immediately, I say again, Climb immediately on heading . . . to . . . feet . . . (further instructions) Acknowledge' |

9 **AREA RADAR**

The procedure whereby pilots may be instructed to 'OMIT POSITION REPORTS' will only be applied when radar identification has been achieved by the Area Radar Controller concerned. The instruction 'RESUME POSITION REPORTING' will be passed before the aircraft is transferred to the next ATCC. See Chapter 5 para 4.2.(b).

# 10 SECONDARY SURVEILLANCE RADAR

## 10.1 Introduction

This section details the phraseology used by controllers at SSR equipped ATSUs to obtain information from transponder equipped aircraft. Pilot phraseology is not given since this is either the aircraft callsign as an acknowledgement or a complete readback of the SSR operating instructions followed by the aircraft callsign.

## 10.2 SSR Phraseology

| *ATSU Phrase* | *Meaning to the Pilot* |
|---|---|
| 'Squawk' | Operate transponder, or check its operating condition. (With SIF set master control to NORMAL) |
| 'Squawk Alpha Code . . . '<br>'Squawk three Code . . . ' | Select the mode and code as applicable: i.e. Mode A (for ICAO SSR equipped aircraft and code . . .<br>Mode 3 (for Military aircraft) and Code . . .<br>(Mode C is normally to be activated at the same time as Modes A or 3) |
| 'Recycle' | Reselect assigned mode and code |
| 'Negative Altimeter | Do not activate Mode C (in conjunction with Modes A or 3) |
| 'Squawk Ident' | Use Identification feature, retain present code. |
| 'Squawk Mayday' | Select A 7700 for Emergency. |
| 'Squawk Standby | Switch to 'standby', retain present code |
| 'Squawk Altimeter' | Switch on Mode C |
| 'Check altimeter setting and confirm level' | Pilot to read back altimeter subscale setting and indicated level for confirmation |

| *ATSU Phrase* | *Meaning to the Pilot* |
|---|---|
| 'Stop altimeter Squawk. Mode Charlie wrong indication' | Stop Mode C transmission. Incorrect level readout |
| 'Stop altimeter Squawk' | Switch off Mode C |
| 'Stop Squawk Alpha' | Switch off Mode A (ICAO) |
| 'Stop Squawk Three' | Switch off Mode 3 (Military) |
| 'Stop Squawk' | Switch off transponder |

## 11 LOWER AIRSPACE RADAR SERVICE

11.1 When within approximately 30nm of a participating aerodrome establish two way RTF communication on the appropriate frequency.

| | |
|---|---|
| *ACFT* | **' . . . (participating aerodrome) this is GBFBO Request Lower Airspace Radar Service'** |

The pilot will be requested to pass his details.

| | |
|---|---|
| *ACFT* | **'Cotswold Radar GBFBO Cessna 310<br>10 miles south of Swindon<br>Heading 340 Flight level 65 IMC** |

11.2 After radar identification the pilot is then expected to:

(a) Follow advice issued by the radar controller.

(b) Maintain a listening watch on the allocated RTF frequency.

(c) Advise the controller before changing heading or flight level.

(d) Advise the controller of any change in flight conditions.

(e) Advise the controller when the service is no longer required.

11.3 Other RTF phraseology used by military controllers is:

| *Situation* | *Phraseology* |
|---|---|
| When the unit or frequency is working to capacity | 'This Unit is working to capacity no service available for . . . minutes' |
| Restriction in radar service | 'You are approaching an area of . . . (Permanent echoes, clutter etc) Will you accept re-routeing?' |

CHAPTER 9

# MISCELLANEOUS PROCEDURES AND PHRASEOLOGY

## 1. METEOROLOGICAL ABBREVIATIONS USED

The following abbreviations are used by the Air Traffic Services in Meteorological Broadcasts. The whole abbreviation is transmitted as a word.

| *Abbreviation* | *Meaning* |
|---|---|
| 'CAVOK' | Visibility, cloud and present weather better than prescribed values or conditions (pronounced KAV-OH-KAY). |
| 'FRONT' | Front, relating to weather. |
| 'GRADU' | Gradual or gradually. |
| 'INTER' | Intermittent. |
| 'NIL' | None. |
| 'NOSIG' | No significant change. |
| 'OBS' | Observe or observation. |
| 'PROB' | Probability. |
| 'RAPID' | Rapid. |
| 'SIGMET' | Information issued by meteorological watch offices concerning certain meteorological phenomena. |
| 'SNOCLO' | Aerodrome unusable due to snow. |
| 'SNOWTAM' | A special series of Notam notifying the presence or removal of hazardous conditions due to snow, ice, slush or standing water associated with snow, slush and ice on the movement area. |

(cont.)

| *Abbreviation* | *Meaning* |
|---|---|
| 'SPECIAL' | Special meteorological report. |
| 'SPOT' | Spot wind. |
| 'TEMPO' | Temporarily. |
| 'TEND' | Trend or tending to. |
| 'TIL' | Until |

## 2 TRANSMISSION OF METEOROLOGICAL INFORMATION TO AIRCRAFT

### 2.1 Introduction

As a general rule controllers will only transmit meteorological information that has been supplied, or agreed by the meteorological office.

(a) Meteorological Reports (ROUTINE)
(b) Meteorological Reports (SPECIAL)
(c) SIGMETS
(d) Forecasts:
  (i) Regional QNH
  (ii) Area and Aerodrome forecasts
(e) Warnings:
  (i) Fog
  (ii) Snow

Exceptions are briefly:

(a) Indicated wind direction and speed when anemometer indicators are fitted in the control room.
  (NB. Indicated wind speed is given in degrees magnetic).
(b) RVR observations.
(c) Sudden deteriorations.
(d) Information received from an aircraft.
(e) Cloud echoes observed on radar (see Chapter 8).
(f) Observations made by controllers who are certified as observers at aerodromes where there is no met office.

(g) 'Unofficial observations' made at aerodromes where no met or certificated staff are available.

In this instance the phraseology will be:
'UNOFFICIAL OBSERVATION FROM . . . (name of aerodrome) GIVES . . . (data)

## 2.2 Automatic Terminal Information Service(ATIS)

(a) *For Departing Flights*

The following met information will be broadcast on published frequencies at aerodromes equipped with ATIS. Pilots need not acknowledge a departure ATIS unless requested on the broadcast.

Runway(s) in use for departures.

Surface wind direction (in degrees magnetic) and speed.

Aerodrome QNH.

Air temperature and dewpoint.

Essential aerodrome information concerning surface conditions.

Details concerning unserviceabilities of departure navigation aids.

Special weather reports, e.g. thunderstorm, hail etc.

Any other information considered to be useful to pilots of departing flights (windshear, marked temperature inversion warnings, etc).

ATIS messages will be consecutively coded starting each day with 'ALPHA'. A new message will be broadcast whenever there is a significant change in any of the items and controllers will pass such changes to pilots on RTF until the new ATIS message is being transmitted.

(b) *For Arriving Flights*

Elements of the following information are broadcast on selected VOR's at aerodromes equipped with ATIS, QFE values are not broadcast. Pilots, on establishment of communication with Approach Control, are to acknowledge receipt of arrival ATIS using the code letter allocated. As a safeguard, if a current code letter is not used or receipt not acknowledged, complete information will be passed, as will those elements omitted:

Runway in use

Current met information and time of observation (including any pilot information on windshear turbulence, thunderstorm, hail etc).

QNH

Current runway surface conditions when appropriate. Changes of status of visual/non-visual aids essential for approach and landing.

## 2.3 Volmet Meteorological Broadcasts

Continuous broadcasts of current meteorological information for certain airports in the United Kingdom and Europe are made by London Area Control Centre using the callsigns LONDON VOLMET, LONDON VOLMET NORTH, LONDON VOLMET SOUTH.

The elements broadcast for each airport are:

Time of observation
Surface wind (in degrees 'true')
Visibility
RVR (if applicable)
Weather
Cloud
Temperature
Dewpoint
QNH

Full details are contained in AIP (Met section).

## 2.4 'Talk to Met' Facility

Some aerodromes have a 'Talk to Met' facility whereby pilots can obtain Met information on RTF outside aerodrome hours. Pilots communicate directly with the Met Office.

## 2.5 Runway Visual Range (RVR)

(a) *Introduction*

There are two methods of RVR observation (Instrumented RVR and human observer).

(b) *Instrumented RVR*

*(i)* RTF transmissions to pilots will refer to the three recording positions:
Touchdown
Mid Point
Stop End

(ii) RVR for the touchdown zone will be given, followed by any other position if the RVR is less than the touchdown value and less than 800 metres.

(iii) If all 3 positions are reported they will be passed to the pilot as three numbers.
e.g. 'RVR 650; 500; 550'
relating to touchdown, mid point, stop end respectively.

(c) *Human Observer Method*

(i) Incremental values up to 1100 metres (exceptionally 1500 metres) will be passed. If the assessed value is more than 1100 metres the controller will say:
'RVR GREATER THAN 1100 METRES'

(ii) If no markers or lights are visible the controller will say:
'RVR LESS THAN 100 METRES'

(iii) Observations from the tower or by pilots of other aircraft may be added to the report, e.g.:
(a) 'RVR . . . METRES THICKER PATCHES FURTHER ALONG THE RUNWAY OBSERVED FROM THE CONTROL TOWER'
(b) 'RVR . . . METRES THICKER PATCHES REPORTED FURTHER ALONG THE RUNWAY BY PILOT OF ARRIVING/DEPARTING AIRCRAFT THAT LANDED/WAS AIRBORNE AT . . . ' (time)

## 2.6 Windshear

(a) *Reports from pilots*

These should include:
A warning of the presence of windshear
Height, or height band, where the shear was encountered.
The time at which it was encountered.

Details of the effect of the windshear on the aircraft, e.g. speed gain or loss, vertical speed tendency, change in drift.

(b) *Report by controllers*

On receipt of windshear information from pilots, controllers will issue details for as long as the conditions are known to exist. Similarly, the local weather situation may necessitate the passing of warnings of potential windshear.

'AT 0915 A DEPARTING B727 REPORTED WINDSHEAR AT 300 FEET. AIRSPEED LOSS 30 KNOTS STRONG LEFT DRIFT.'

2.7 **Marked Temperature Inversion**

A warning transmitted either on the departure ATIS or by the controller, for the information of departing aircraft.

'A MARKED TEMPERATURE INVERSION IS PRESENT FROM THE SURFACE TO . . . FEET AND THE TEMPERATURE DIFFERENCE BETWEEN THE BOTTOM AND THE TOP OF THE LAYER IS LIKELY TO BE 10°C OR MORE.'

3. **AERODROME OPERATING MINIMA**

Recommended operating minima for non public transport flights carrying out instrument approach procedures are passed by ATC when the lowest reported cloud is at, or below the visual manoeuvring height or the Met visibility is 1500 metres or less.

'THE RECOMMENDED MINIMA FOR THIS APPROACH ARE DECISION HEIGHT . . . FEET AND VISIBILITY . . . METRES.'

4. **RUNWAY BRAKING ACTION**

4.1 **Reports by pilots**

During adverse weather conditions of snow/rain it is extremely helpful if pilots pass details of braking action to controllers so that possible remedial action may be taken and other pilots warned.

4.2 **Reports by Controllers**

(a) *Wet runways*

(i) 'BRAKING ACTION GOOD 0.55'

(ii) 'BRAKING ACTION MEDIUM 0.43 HEAVY RAIN TIME OF MEASUREMENT 0830'

(iii) 'BRAKING ACTION POOR 0.43, 0.35, 0.39 STANDING WATER. TIME OF MEASUREMENT 1440'

(b) *Wintry conditions*

(i) BRAKING ACTION MEDIUM SLUSH AT 0940 (time)

(ii) BRAKING ACTION POOR ICE AT 1020 (time)

4.3 **Interpretation**

The full procedures and interpretation of braking action are detailed in the AIP (AGA section).

5 **AIRMISS REPORTING BY RTF**

An Airmiss Report should be made by any pilot flying in the United Kingdom Flight Information Region, the Upper Flight Information Region or Shanwick Oceanic Area when he considers that a risk of collision has been occasioned during flight by the proximity of another aircraft.

The initial report is made by RTF to the ATSU in communication with the aircraft except that if the controllers workload is such that he is not able to accept the report the pilot will be requested to file details after landing.

The Pilot's RTF report should commence with 'AIRMISS REPORT'

'Airmiss Report
Aircraft callsign
Position of airmiss
Aircraft heading
Flight level, altitude or height
Altimeter setting
Aircraft attitude (Level/climbing/descending/turning)
Weather conditions

Date and time (GMT) of the Airmiss
Description of other aircraft
First sighting distance and details of flight paths of reporting and reported aircraft'.

RTF Airmiss reports are confirmed in writing within seven days of the incident to allow follow up action to be taken. AIP (RAC Section.)

## 6 DIVERSIONS

If an aircraft operating agency proposes to divert one of its aircraft the message may be passed by an ATSU.

| | |
|---|---|
| *ATSU* | 'Company Advise Divert to . . . (aerodrome)<br><br>Weather at . . . (diversion aerodrome)<br><br>Reason for Diversion . . .<br><br>. . . (clearance instructions)<br><br>Acknowledge' |

Exceptionally ATC may advise a pilot to divert before consulting the aircraft operating agency.

| | |
|---|---|
| *ATSU* | 'Request Divert to . . . (aerodrome)<br><br>Weather at . . . (diversion aerodrome)<br><br>Reason for Diversion . . .<br><br>. . . (clearance instructions)<br><br>Acknowledge' |

In both cases, if unable to comply, the pilot should give his reasons and state his intentions.

## 7 OIL POLLUTION REPORTING BY RTF

Pilots sighting substantial patches of oil are requested to make reports by RTF to the ATSU with whom they are in communication, or the appropriate FIS in order that action can be taken.

The RTF reports should contain the following:

'OIL POLLUTION REPORT'

. . . Time and (date if required) pollution was observed,

. . . Position and extent of oil slick,

. . . Name and nationality, or description, including any distinctive markings, or any vessel seen discharging oil.

Information on the following may also be included.

(a) Assessment of the course and speed of any vessel seen discharging oil.

(b) Whether any oil was observed ahead of the discharging ship, and the estimated length of the slick in her wake.

(c) The direction in which the oil was drifting.

(d) The identity of any other vessels in the immediate vicinity.

## 8 INTERCEPTIONS BY MILITARY AIRCRAFT—RTF

Pilots are warned that should they become involved in an interception by military aircraft they should, as soon as possible:

(a) Notify the appropriate air traffic service unit.

(b) Attempt to establish radio communication with the intercepting aircraft by making a general call on the emergency frequency of 121.5 MHz and repeating the call on 243 MHz if practicable and using the callsigns 'INTERCEPTOR' and 'INTERCEPTED AIRCRAFT'.

The identity of the aircraft and the nature of the flight should be included.

(c) Select Mode A Code 7700 if equipped with SSR unless instructed otherwise by the appropriate ATC Unit.

| | |
|---|---|
| *ACFT* | **'Interceptor this is Intercepted Aircraft my Callsign is GBFBO on an IFR Flight to . . . (destination) In Communication with . . . ' (ATSU).** |

If any instructions received by radio from any source conflict with those given visually or by radio from the intercepting aircraft, the intercepted aircraft shall request immediate clarification while continuing to comply with the instructions given by the intercepting aircraft.

# CHAPTER 10

# MISCELLANEOUS INFORMATION

## 1 RADIO EMISSIONS

### 1.1 Classification

Radio emissions are classified according to the following characteristics:

(a) Type of modulation of main carrier.

(b) Type of transmission.

(c) Supplementary characteristics.

Sub divisions of (a), (b) and (c) are allocated alpha-numeric symbols as shown in the following tables and then grouped in sequence to provide a symbolic designation for the transmission.

(a) *Type of Modulation of Main Carrier* *Symbol*

| Type of Modulation of Main Carrier | Symbol |
|---|---|
| Amplitude | A |
| Frequency (or Phase) | F |
| Pulse | P |

(b) *Type of Transmission* *Symbol*

| Type of Transmission | Symbol |
|---|---|
| Absence of any modulation intended to carry information | 0 |
| Telegraphy without the use of a modulating audio frequency | 1 |
| Telegraphy by the ON/OFF keying of a modulating audio frequency or audio frequencies, or by the ON/OFF keying of the modulated emission (special case: an unkeyed modulated emission) | 2 |
| Telephony (including sound broadcasting) | 3 |
| Facsimile (with modulation of main carrier either directly or by a frequency modulated sub-carrier) | 4 |

(b) *Type of Transmission* (cont.) *Symbol*

| Type of Transmission | Symbol |
|---|---|
| Television (vision only) | 5 |
| Four frequency duplex | 6 |
| Multi-channel voice frequency telegraphy | 7 |
| Cases not covered by the above | 9 |

(c) *Supplementary Characteristics* *Symbol*

| Supplementary Characteristics | Symbol |
|---|---|
| Double sideband | (none) |
| Single sideband: | |
| Reduced carrier | A |
| Full carrier | H |
| Suppressed carrier | J |
| Two independent sidebands | B |
| Vestigial sideband | C |
| Pulse: | |
| Amplitude modulated | D |
| Width (or duration) modulated | E |
| Phase (or position) modulated | F |
| Code modulated | G |

1.2 **Examples**

The following classes of emissions are used in the aeronautical mobile and radionavigation services.

| *Facility* | *Class of emission* |
|---|---|
| ILS and Marker beacons | A2 |
| Communication by radiotelephony—VHF | A3 |
| —HF | A3<br>A3H<br>A3J |
| Long Range NDB's (class AO emission with an identification signal of A1 emission), | A0A1 |
| Short and medium range NDB's (Class A0 emission with an identification signal of A2 emission), | A0A2 |
| VOR's | A9 |
| SSR, DME, TACAN, LORAN C | P |

VHF RTF is used for all air-ground communication throughout the airspace under UK jurisdiction except that HF (A3, A3H and A3J) is also used in the Shanwick Oceanic Control Area and UHF is available at the London ACC and at certain aerodromes.

Specific emissions used at aeronautical stations are shown in Column 4 (EM) of the UK Air Pilot (Radio Communication and Navigation Facilities) beginning COM2-1.

## 2 FREQUENCY SUB-DIVISION

Frequencies are normally expressed:

(a) in Kilohertz (KHz) up to 3000 KHz
(b) in Megahertz (MHz) thereafter to 3000 MHz
(c) in Gigahertz (GHz) thereafter to 3000 GHz

| *Frequency sub division* | *Range* |
|---|---|
| VLF—Very Low Frequency | 3-30 KHz |
| LF—Low Frequency | 30-300 KHz |
| MF—Medium Frequency | 300-3000 KHz |
| HF—High Frequency | 3-30 MHz |
| VHF—Very High Frequency | 30-300 MHz |
| UHF—Ultra High Frequency | 300-3000 MHz |
| SHF—Super High Frequency | 3-30 GHz |
| EHF—Extremely High Frequency | 30-300 GHz |
| | 300-3000 GHz |

Note. KHz = Hz × 1000
MHz = KHz × 1000
GHz = MHz × 1000
1 Hz = 1 cycle per second

## 3 SYLLABUS

Syllabus for test and examination for the grant of the Flight Radiotelephony Operator's licence (*extract from CAP 90 as amended*).

### 3.1 Section A—Practical Communication Tests

Candidates will be required to have a knowledge of standard

phraseology and procedures and to carry out typical RTF communications, including the following:

use of RTF spelling alphabet as prescribed in CAP 413 Radiotelephony Procedures and Phraseology;

obtaining all required information prior to departure;

pre-flight radio test transmissions;

taxi, take-off, and airways or other appropriate route clearance;

reporting point procedures;

obtaining VHF DF assistance;

distress and urgency procedures;

recording of VHF weather broadcasts;

obtaining meteorological information on a 'request' basis;

entering Control and Military Air Traffic Zones.

The test will be conducted using a synthetic RTF circuit. Headphones, microphones and operating controls will be supplied. During the test engine noise may be simulated and where appropriate radio interference may also be injected into the circuit. The speech quality and strength, interference and general method of conducting the test will be arranged to represent actual working conditions as far as possible. The candidate will be required to demonstrate his knowledge of and ability to make rapid changes of frequency.

3.2 **Section B—Procedure and Regulations (Written paper)**

The examination will include questions on the following:

definitions of Air Traffic Services;

standard phraseology used in radiotelephony communications;

order of priority of communications;

procedure during distress traffic and conditions justifying transmission of distress and urgency signals;

brevity of communications;

classification and types of DF bearings and positions;

knowledge of publications and notices (eg CAP 413, Aeronautical Information Circulars etc) in which information required under this section is promulgated.

## 4 PILOT COMPLAINTS CONCERNING AERONAUTICAL TELECOMMUNICATIONS

Pilot reports of faults concerning services and facilities in the Aeronautical Mobile, Broadcast and Navigation services may be recorded on the CAA Form CA647. The pilot should ensure that the Briefing Office, STO or ATCO at the destination or Airport of first landing receives full details in order that remedial action can be taken. Reports of local unserviceabilities will be forwarded to the Telecommunications staff if received on RTF by the ATSU.

## 5 AIR TRAFFIC SERVICE COMPLAINTS ABOUT PILOT COMMUNICATIONS

Aircraft radio faults including technical failure, incorrect operating procedures and misuse of specific radio channels may result in the aircraft operator receiving a CAA Form CA163 which details the fault condition and invites the operator to explain and/or state what corrective action has been taken.

*APPENDIX A*

# MESSAGE CATEGORIES

## 1 FLIGHT SAFETY MESSAGES

1.1 Movement and control messages

(a) Flight Plan

(b) Amendment and co-ordination. To include:

- (i) Departure
- (ii) Delay
- (iii) Arrival
- (iv) Boundary estimate
- (v) Modification
- (vi) Co-ordination
- (vii) Acceptance

(c) Cancellation

(d) Clearance

(e) Transfer of control

(f) Request

(g) Position report

1.2 Messages originated by the operating agency of immediate concern to an aircraft in flight or to an aircraft about to depart.

1.3 Meteorological advice of immediate concern to aircraft in flight or about to depart (individually communicated or broadcast).

1.4 Other messages concerning aircraft in flight or about to depart.

## 2 METEOROLOGICAL MESSAGES

2.1 Meteorological forecasts.

2.2 Meteorological observations.

2.3 Meteorological messages exchanged between meteorological offices.

## 3 FLIGHT REGULARITY MESSAGES

3.1 Load messages.

3.2 Aircraft operating schedules.

3.3 Aircraft servicing.

3.4 Changes in collective requirements for passengers crew or cargo.

3.5 Non routine landings to be made by aircraft en-route or about to depart.

3.6 Parts and materials urgently required.

3.7 Preflight arrangements of air navigation services and operational servicing.

3.8 Aircraft operating agency messages relating to aircraft arrival and departure.

## *APPENDIX B*

EXTRACT FROM ANNEX 10 VOL II ATTACHMENT B—development of radiotelephony speech for international aviation

### BASIC PRINCIPLES

(a) The English language should be the basis for the development of the requisite phraseology. Words with Latin roots should be given preference in developing phraseology.

(b) Words and phrases should be selected in such a way as to ensure optimum transmissibility over radiotelephone channels and should be incapable of misinterpretation.

(c) Words and phrases should be avoided which will be liable to difference of pronunciation likely to cause misunderstanding.

(d) Spoken Q code groups, which by their common usage, have already become part of aviation terminology, may be used where they provide a preferable alternative to a long or complex phrase eg QFE, QFF, QNE, QNH, QTE.

(e) Where phrases already in general use have proved by experience to be phonetically suitable irrespective of the language from which they were derived, they should not be arbitrarily changed.

(f) New phraseology developed during the study should be clear, unambiguous and, where practicable, concise. However, clarity should not be sacrificed in the interest of brevity.

(g) Phrases should be developed on the principle that they represent a thought expressed in a live language; however, the grammatical construction should be as simple as possible.

(h) Positive and negative instructions or advice should be clearly differentiated.

(i) Where practicable words containing sounds or syllabic constructions traditionally difficult in pronunciation by non-English speaking personnel should be avoided.

*APPENDIX C*

# RELEVANT LEGISLATION

## AIR NAVIGATION ORDER

Article 14 Radio Equipment of aircraft

Schedule 6 Radio Equipment to be carried in aircraft

Article 35 Operation of radio in aircraft

69 Radio equipment at Aerodromes

70 Records at aerodromes

## RULES OF THE AIR AND AIR TRAFFIC CONTROL REGULATIONS

Rule 4 Reporting Hazardous Conditions
16 Weather Reports and Forecasts
20 Notification of Arrival and Departure
27 Flight Plan and Air Traffic Control Clearance
28 Position Reports
34 Flight within Aerodrome Traffic Zones
35 Use of Radio Navigation Aids
47 Distress, Urgency Signals

# INDEX

# NOTES

# NOTES